Biological Variation

Biological Variation
From Principles to Practice

Callum G. Fraser, PhD
Ninewells Hospital and Medical School
Dundee Scotland

2101 L Street, NW, Suite 202
Washington, DC 20037-1558

©2001 American Association for Clinical Chemistry, Inc. All rights reserved. No part of this publication may be reproduced, stored in a retrieval systems, or transmitted in any form by electronic, mechanical, photocopying, or any other means without written permission of the publisher.

4 5 6 7 8 9 0 V G I X 03

Printed in the United States of America

Library of Congress Cataloging-in-Publication Data

Biological variation : from principles to practice / by Callum G. Fraser.
 p. ; cm.
 Includes index.
 ISBN 1-890883-49-2 (alk. paper)
  1. Clinical chemistry. 2. Reference values (Medicine) 3. Variation (Biology) I. Fraser, C. G. (Callum G.)
  [DNLM: 1. Reference Values. 2. Clinical Laboratory Techniques—standards. 3. Reproducibility of Results. QY 16 B615 2001]
RB40 .B57 2001
616.07′56—dc21   2001035550

Contents

Preface		vii
Foreword		ix
Chapter 1	The Nature of Biological Variation	1
Chapter 2	Quality Specifications	29
Chapter 3	Changes in Serial Results	67
Chapter 4	The Utility of Population-Based Reference Values	91
Chapter 5	Other Uses of Data on Biological Variation	117
Chapter 6	Next Steps	123
Glossary		129
Appendix 1	Data on Components of Biological Variation	133
Appendix 2	Quality Specifications for Precision, Bias, and Total Error Allowable	139
Index		145

Preface

This book provides a simple but comprehensive guide on how to

- produce numerical data for within-subject and between-subject components of biological variation, or
- locate these in available literature or other sources, and then
- use the information in everyday clinical laboratory practice.

There is significant renewed interest in strategies for setting quality specifications as well as for setting population-based and subject-specific reference values. Up to date relevant information is not readily available in other texts but is scattered throughout the literature. That information has been collected in this book.

The intended readership includes all those who are concerned with quality in laboratory medicine, which should embrace all who work in laboratories that process patient specimens, and especially

- laboratory professionals concerned with quality administration and management,
- those undertaking undergraduate and postgraduate courses in laboratory medicine, and
- trainees who are taking professional examinations or seeking certification.

I hope that much of the content will interest those who request laboratory tests and interpret numerical results, and that this content will be used to educate medical students in these often less than well taught topics.

In some ways this book updates my previous concise volume, *Interpretation of Clinical Chemistry Laboratory Data* (Blackwell Scientific Publications Ltd., Oxford, 1986). However, the content and emphasis have changed considerably.

The impetus to writing the book resulted from a meeting in early 2000 with Zoe Brooks, President of QIK Quality Is Key, Ltd., of Worthington, Ontario, Canada. Zoe visited me in Dundee to see how we actually used the models and data that we had produced and published over the last 20 years. Although I had resisted all previous attempts to update my earlier book, she enthused me enough to document current thoughts on how to translate biological variation from principles to practice.

This book is the result of our close collaboration. I wrote the first draft of the text. Zoe read every word and added many ideas to this, the final version. Jeanette Devereaux drew all the diagrams and put up with picky comments on her excellent drafts. AACC Press translated my writings into a more concise and readable format. I am, of course, responsible for all errors and omissions.

                                      Callum G. Fraser
                                      Ninewells Hospital and Medical School
                                      Dundee
                                      Scotland
                                      February 2001

Foreword

Many analytes of interest in the clinical laboratory can vary over an individual's lifetime, simply because of natural biological factors involved in the aging process. These variations may occur rapidly at critical points in the life cycle, such as during the neonatal period, childhood, puberty, menopause, or old age.

In addition, certain analytes have predictable biological rhythms or cycles. These cycles may be daily, monthly, or seasonal. Knowledge of these cycles is vital for good patient care. For example, a patient sample must be collected at the time in the cycle that is appropriate for the clinical purpose to which the test result will be applied. Since developing good reference values is complex and time consuming, it is important to generate these values correctly, particularly at clinically important decision-making points. Moreover, the absence of an expected rhythm or cycle can give important clues about the presence of disease and is the simplest of dynamic function tests.

Most analytes, however, do not have cyclical rhythms that are of major clinical importance. In fact, the variation can be described as random fluctuation around a homeostatic setting point. We see this easily in practice. If we take a series of samples from one individual for a particular laboratory test, then the results are not all exactly the same number. The test results of any person vary over time, due to three factors:

- **pre-analytical influences**—those related to preparation of the individual for sampling, such as posture; and those influenced by sample collection itself, such as tourniquet application time,
- **analytical random error** (precision)—and possibly **systematic error** (changes in bias due to calibration, for example), and
- **inherent biological variation** around the homeostatic setting point (this is called *within-subject* [or *intra-individual*] biological variation).

If we performed the same test on various individuals, we would find that the mean of each person's results would not all be exactly the same number. Individual homeostatic setting points usually vary. This difference between individuals is called *between-subject* (or *inter-individual*) biological variation.

In order to determine the magnitude of within-subject and between-subject components of biological variation in numerical terms, we could conduct a rather simple experiment along the following lines.

1. Recruit a small group of apparently healthy volunteers (or patients without any disease that affected the analyte under investigation).

2. Take a series of samples from each individual at regular intervals while minimizing pre-analytical variation.
3. Store the samples for analysis.
4. Perform the analysis in duplicate while minimizing analytical sources of variation.
5. Remove outliers, that is, numbers that are different from the bulk of the data set.
6. Determine the analytical, and within-subject and between-subject biological components of variation using simple statistical analysis of variance techniques (ANOVA).

APPLICATION OF DATA ON RANDOM BIOLOGICAL VARIATION

If we generated numerical data using this experimental technique, we would then have quantitative knowledge of

- average within-subject biological variation and
- between-subject biological variation.

We rarely do this type of experimental work in our own laboratory because the literature contains large databases about the components of biological variation. These are easy to access. It is generally appropriate to use these data in everyday practice.

Data on the components of biological variation can be applied to set quality specifications for

- precision,
- bias,
- total error allowable,
- the allowable difference between methods,
- use in proficiency testing programs (PT) or external quality assessment schemes (EQAS), and
- reference methods.

Data on within-subject biological variation and analytical precision can be used to

- determine the change that must occur in an individual's serial results before the change is significant (the reference change value),
- determine the statistical probability that a change in an individual's serial results is significant, and
- generate objective delta-check values for use in quality management.

Comparing within-subject and between-subject biological variation allows us to

- decide the utility of traditional population-based reference values (often termed normal ranges), and
- clarify why stratifying reference values according to age and sex, for example, improves clinical decision making.

Data on biological variation can be used for other purposes, including

- calculating the reliability coefficient used in epidemiology,
- determining the number of samples needed to get an estimate of the homeostatic setting point within a certain percentage with a stated probability, and
- deciding the best way to report test results, the best sample to collect, and the test procedure of greatest potential use.

And, of course, generation and application of data on biological variation is an essential prerequisite in the evolution of any new test procedure.

BIOLOGICAL VARIATION AND THE QUEST FOR QUALITY

Over time, we have come to realize that quality management involves much more than the simple statistical quality control techniques that we have performed every day at the bench for many years—it requires incorporating and integrating quality laboratory practice, quality assurance, quality improvement, and quality planning as well as quality control. In short, quality management impacts all phases of obtaining a clinically appropriate and correctly interpreted laboratory result, including the pre-analytical, analytical, and post-analytical phases of our work.

It is thus vital for those concerned with quality management to know how to generate or find, and then apply, numerical data on the components of biological variation in their everyday practice. Importantly, one must consider the influence of biological variation on laboratory tests and on the interpretation of laboratory results.

DEFINING QUALITY

The International Organization for Standardization (ISO) defines quality as "the totality of characteristics of an entity that bear on its ability to satisfy stated and implied needs." This rather complex definition can be translated to mean—at least for us—that the quality of tests performed in laboratory medicine must allow our clinicians to practice good medicine.

Before we can control, practice, assure, or improve laboratory quality, we must know exactly what level of quality we need to ensure satisfactory clinical decision making. And, since a laboratory service includes much more than the technical analysis of samples, we must appreciate that the time of day or month when patient samples are obtained may influence the test result, and that biological variation influences the interpretation of numerical test results both in monitoring, in which serial results from an individual are assessed for change, and in diagnosis and case-finding, in which population-based reference values are most often used.

USING THIS BOOK

This book brings together modern and recent concepts on the generation and application of data on biological variation. Specifically, this book will help you

- appreciate the many sources of variation in laboratory test results,
- understand the clinical importance of cyclical biological rhythms,
- generate and/or find quantitative data on within-subject and between-subject biological variation,
- appreciate the hierarchy of models available to set quality specifications,
- set quality specifications for precision and bias using biological variation data,
- develop quality specifications for total error allowable based on biological variation,
- recognize that quality specifications for other uses can be generated from biological variation data,
- appreciate the reasons for changes in serial results in an individual over time,
- calculate reference change values for use in monitoring individuals over time,
- calculate the probability that a change in serial results has occurred in an individual,
- calculate delta-check values for use in laboratory quality management using patient data,
- generate population-based reference values and stratify (partition) these where appropriate,
- understand the limitations of conventional reference values due to biological individuality,
- appreciate the many other minor uses of data on biological variation, and
- recognize that data on biological variation are essential prerequisites to introducing new tests.

This real aim of this book is to describe the generation and the many applications of quantitative data on random biological variation in many facets of laboratory medicine. In addition, we will explore easily available sources of these data to make the applications possible everywhere.

Chapter 1
The Nature of Biological Variation

Healthcare consumers rarely have only one set of laboratory results in their case records. Even casual inspection of serial medical laboratory data on any individual will show that laboratory test results vary over time. In fact, variation in test results is expected. This chapter presents an overview of how and why test results vary and briefly examines sources of pre-analytical, analytical, and biological variation.

A BIOLOGICAL (AND PERSONAL) CASE HISTORY

I consider myself fairly healthy. Because I am interested in biological variation and relatively keen on the promotion of health and the prevention of disease, I have performed a variety of laboratory tests on myself at fairly regular intervals over the years. A few years ago, at 53 years of age, I requested a group of laboratory medicine tests, particularly those that investigate renal function, liver function, lipids, thyroid function, and prostate specific antigen (PSA).

The results of some of these tests are shown in Table 1.1. No unusual results appeared, except for serum bilirubin concentration, which exceeded the upper reference limit. In previous years, I had always disregarded this because of the known pitfalls of population-based reference values—which we will explore later in this book.

As a clinical scientist, however, and in part because of the research interest of a colleague, I became interested in exploring this "unusual" finding further. After one month, I repeated the panel of biochemical tests. The second set of results investigated the "abnormal" bilirubin and was done to "confirm" the other "normal" test results because that is the approach that many clinicians would take. Note the following.

- The serum bilirubin concentration is not identical over time but does remain unusual and above the upper reference limit. (Further investigation using new molecular genetic techniques showed that, as I had always suspected but was now able to prove, I have the common and harmless Gilbert's syndrome, which fully explains the high serum bilirubin concentration.)
- Results for all other tests still lie within the appropriate reference intervals—but the actual numbers reported by the laboratory are not the same.

Perhaps the more important questions are these.

- Why do numbers reported by the laboratory differ over time?
- Are the apparent rises and falls in numerical results significant?
- Do the changes in numbers reflect signs of early or latent pathology?

Table 1.1 Test Results, 53-Year-Old Male

Analyte	1st Result	Units	Reference Interval	2nd Result
Sodium	139	mmol/L	135–147	139
Potassium	4.3	mmol/L	3.5–5.0	4.1
Urea	4.0	mmol/L	3.3–6.6	4.4
Creatinine	88	µmol/L	64–120	97
ALT	40	U/L	12–40	28
Bilirubin	19	µmol/L	0–17	21
Alk Phos	49	U/L	30–105	46
GGT	57	U/L	11–82	49
Calcium	2.39	mmol/L	2.10–2.55	2.33
Albumin	44	g/L	35–50	48
Cholesterol	4.60	mmol/L	Ideal < 5.00	4.82
Triglycerides	0.48	mmol/L	Up to 2.30	0.52
TSH	2.03	mU/L	0.4–4.0	2.19
PSA	1.5	µg/L	Up to 4.0	2.5

SOURCES OF TEST RESULT VARIATION

As it happens, test results do inherently vary. Here we explore the three sources of variation:

- pre-analytical variation,
- analytical variation (precision and changes in bias), and
- within-subject biological variation.

PRE-ANALYTICAL VARIATION

Patient Preparation   Before any sample is ready for analysis in the medical laboratory, the individual must be prepared for sample collection. Then the sample must be collected, transported to the laboratory, and uniquely identified, handled, and sometimes stored, prior to the actual analysis.

Each of these processes may vary and therefore affect the pre-analytical phase of laboratory testing. It is impossible to describe all of these effects in quantitative terms in this book. However, certain variables have been found to affect samples collected for testing (see Table 1.2).

Sample Collection and Handling   In the ideal world, the sources of variation relevant to interpretation of test results are carefully documented before samples are collected and transported to the laboratory, identified, handled, and sometimes stored. Not only does each stage in the processes between collection and analysis have inherent sources of variation, but the possibilities for variation in this second

THE NATURE OF BIOLOGICAL VARIATION

Table 1.2 Variables That Affect Samples Collected for Testing

Fasting	If the subject has recently eaten, serum triglycerides, aspartate aminotransferase, bilirubin, glucose, phosphate, potassium, and alanine aminotransferase will rise, generally by more than 5%.
Starvation	Prolonged starvation will decrease serum proteins, cholesterol, triglycerides, and urea, and will increase urate and creatinine.
Exercise	Prolonged exercise raises the serum activity of enzymes (including creatine kinase, lactate dehydrogenase, and aspartate aminotransferase) found in muscle cells.
Altitude	Adaptation to high altitude (which is a long process, involving weeks) increases serum CRP and urate, and hemoglobin and hematocrit.
Stimulants	Caffeine, nicotine, alcohol, and other drugs of misuse affect a number of commonly requested analytes.
Posture	Changes in posture affect • large molecules, such as total proteins; • enzymes; • what appear at first sight to be small molecules but which, in reality, are large molecules that are in whole or in part protein-bound, such as calcium (half bound), iron, steroid and thyroid hormones; and • cells. Samples collected while subjects are standing measure about 10% higher than those collected while subjects are lying down. Samples collected while subjects are seated have intermediate values (see Figure 1.1).

component of the pre-analytical phase (collection) are also considerable. Some important considerations are shown in Table 1.3.

ANALYTICAL SOURCES OF VARIATION

Every analytical or measurement technique has some intrinsic sources of variability. Though these cannot be totally eliminated, they can be minimized by quality laboratory practice and judicious selection of good methodology.

Traditionally, variation is of two types—random and systematic—and these are usually termed precision and bias respectively. (Some purists do not like these terms but they are widely applied and understood and thus will be used exclusively throughout this book.)

Random Variation The International Organization for Standardization (ISO) defines precision as "the closeness of agreement between independent results of measurements obtained under stipulated conditions." In practice, precision is measured by replicate analysis of the same sample. Note that the precision found may be

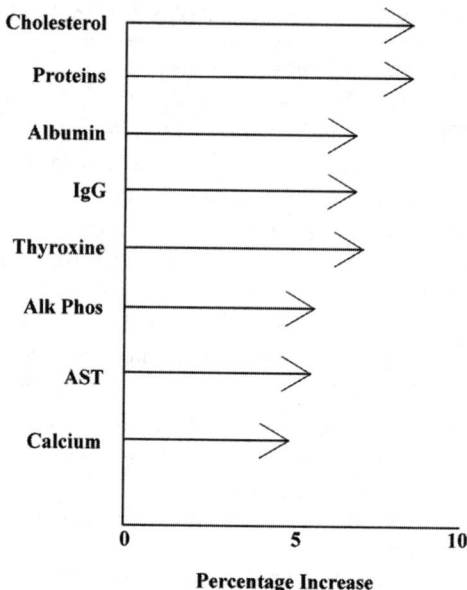

Figure 1.1 Percentage Increase in Some Analytes Upon Changing Posture (from Lying Down to Standing Up)

Changes can be explained by water shifts in the vascular space when subjects move from recumbent to upright positions.

influenced by the analytical conditions. If we took replicate measurements over a short time scale with one operator, one set of reagents, and one calibration, we would likely generate a much smaller precision than if we took replicate results over longer time periods with many operators, different lots of reagents, and different calibrations.

Since random measurement variation is by definition Gaussian, the distribution of results is symmetrical and bell-shaped. Thus, the width of the distribution (better called the dispersion of the distribution), can be calculated as the standard deviation (SD). If one wishes to derive the standard deviation relative to the mean, the coefficient of variation (CV) can be calculated as (SD/mean) * 100. The properties of the distribution are such that it is simple to calculate the percentage of values encompassed by the mean ± n*SD as shown in Figure 1.3.

Random variation is inherent to the analytical system and methodology used, arising from sources such as fluctuations in temperature, variability in volume of sample and/or reagent delivered by pipette or dilutor, changes in environment, and inconsistent handling of materials.

If a method has good precision, its random variation will be low, and the results obtained with such a method will not change much over time—at least due to analytical influences. In contrast, if a method has poor precision, the large analytical random effects may confound many important clinical changes. For instance,

THE NATURE OF BIOLOGICAL VARIATION

Table 1.3 Potential Sources of Variation in Sample Collection

Type of sample	Capillary blood and venous blood are not identical, particularly with regard to glucose: pO_2 partial pressures differ in capillary and arterial samples.
Anticoagulant	Serum and plasma are not identical for all constituents as is sometimes supposed. Proteins, albumin, and transferrin are higher in plasma. Potassium, lactate dehydrogenase, and phosphate are higher in serum, probably due to clot formation and retraction.
Tourniquet application	Prolonged tourniquet application forces small molecules and water out of the intra-vascular space, leaving large molecules, small molecules bound to large molecules (such as bilirubin, and calcium, which is about 50% bound), and cells. Thus plasma becomes more concentrated. In consequence, higher values are found for such analytes after stasis (see Figure 1.2).
Transport time	Glucose in samples that have not been preserved will fall rapidly; serum potassium, phosphate, aspartate aminotransferase, and lactate dehydrogenase will rise if samples remain in prolonged storage prior to centrifugation. (Potassium actually falls in the short term if samples are kept refrigerated.)
Centrifugation	Samples centrifuged for a very short time may contain cellular elements in what is supposedly plasma or serum: artefactually high enzyme activities and potassium can occur. Rapid centrifugation of very fresh samples is sometimes required for unstable analytes such as homocysteine.
Storage	Samples must be stored correctly prior to analysis. Correct storage is a complex process and many considerations must be taken into account—for example, samples for bilirubin assay must be kept out of the light, contact with air must be minimized to prevent general concentration and other effects such as loss of CO_2, and some samples must be frozen rapidly after separation to ensure analyte stability.

we may not be able to see the clinically important "signal" because of all the analytical "noise." This topic will be explored in detail later in this book.

Systematic Variation (Bias) The ISO defines bias as "the difference between the expectation of measurement results and the true value of the measured quantity." In practice, bias is the difference between the results we obtain and some estimate of the true value.

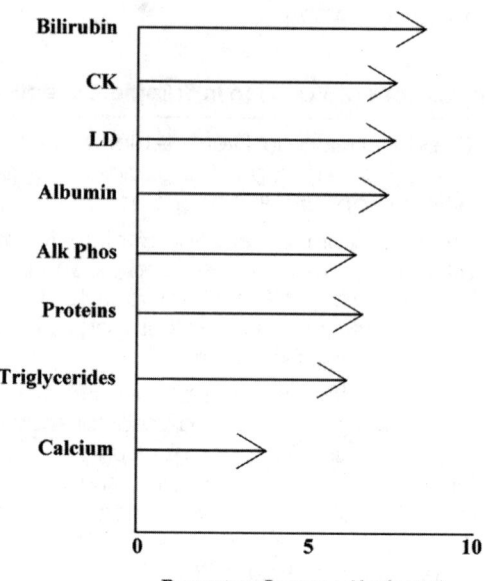

Figure 1.2 Percentage Increase in Some Analytes after Prolonged Venous Stasis

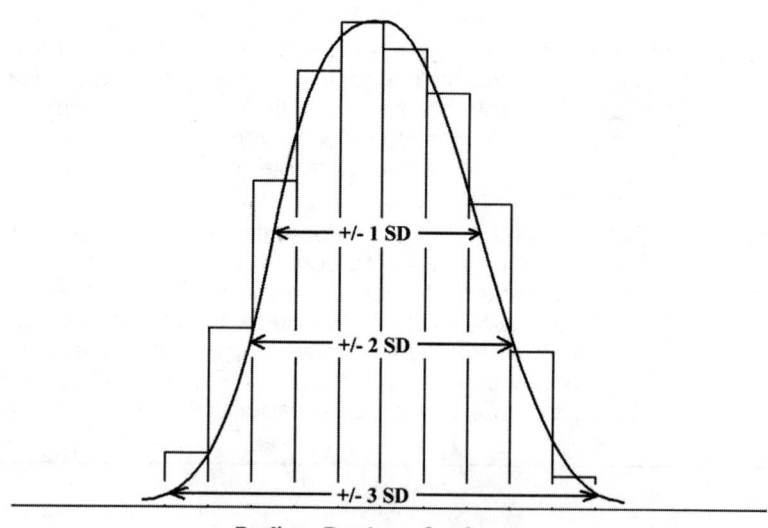

Figure 1.3 Characteristics of the Distribution Found on Replicate Analyses of a Sample

The characteristics of the Gaussian distribution are that
- mean ± 1SD encompasses about 68.3% of the values,
- mean ± 2SD encompasses about 95.5% of the values, and
- mean ± 3SD encompasses about 99.7% of the values.

THE NATURE OF BIOLOGICAL VARIATION

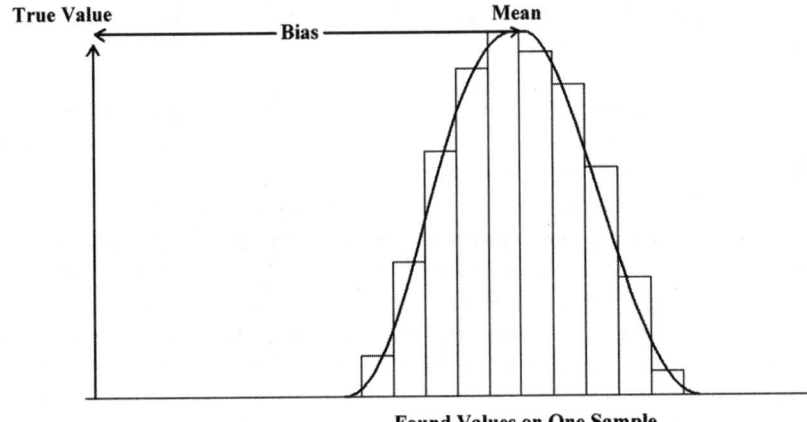

Figure 1.4 Bias—The Difference Between Results and the "True" Value

Estimates of bias can be taken as the difference between values obtained by the laboratory for the samples circulated in PT or EQAS challenges. PT and EQAS organizers determine how laboratory results are compared in terms of values. For example, results may be compared with consensus values, reference method values, or values set by a group of better (reference) laboratories. The concept of bias is illustrated in Figure 1.4.

Bias is a major issue when test results are used in diagnosis, case finding, or screening. In these clinical settings, we often compare test results with pre-set numerical criteria. However, if the laboratory has a constant bias, then clearly systematic variation will not be a problem in the clinical setting of monitoring—at least when we are assessing changes in an individual over time. Test results may all be higher or lower than the true values but none of the variation over time will be attributable to bias.

Note that bias is not always constant. If we make major changes to instrumentation or methodology, and if samples from an individual are compared before and after such changes, differences in bias may contribute to the test result variability seen over time. Moreover, if we use different instruments or assay systems, for example, for routine and emergency (STAT) samples, or in laboratory and point-of-care testing (POCT) or alternate site settings, then these may have different magnitudes of bias.

Our strategy should be to eliminate known biases before reporting analytical results. Moreover, when different assay systems have different biases, consider one system the "gold standard" and calibrate other systems against it.

Over the short- and medium-term, the potential to change the bias of a single method exists when we

- change calibrator lots,
- change reagent lots of reagents and the lots of other consumables, or

- change operators—who, as individuals, may perform tasks very consistently but rather differently from others.

On traditional internal quality control charts, these changes manifest as systematic shifts in the mean, while at the same time, the short term SD or CV stay much the same (see Figure 1.5).

There are four instances when replicate analyses have a very narrow dispersion (i.e., when the assays have good intrinsic precision and a small SD). However, these four instances represent times following four new calibrations. Assay bias changes significantly at each calibration. With many modern analytical systems, changes in bias on calibration are far greater than the inherent analytical random variation—a factor that becomes very important in modern laboratory analytical quality control and management.

However, if we calculate SD over the period of time during which these shifts occur, the found "precision" will include these types of systematic variation. Thus, for most purposes of objectively considering changes in an individual over time, we can consider systematic variation to be negligible, although we should try hard to minimize any systematic shifts in results over time through adopting quality laboratory management techniques.

The most important strategy to minimize bias over time involves improving quality management when re-calibration is performed. The full detail of how to do this falls outside the scope of this book but, in general terms, one should assay multiple quality control samples and then apply rather rigid quality control rules for acceptance of results immediately after re-calibration. It is not necessary to have only one set of control rules and simply assay a small number of control samples as routine practice; in fact, the rules and the number analysed after calibration can differ (to ensure only small changes in bias have occurred) compared to routine assays (to ensure that both precision and bias meet quality specifications).

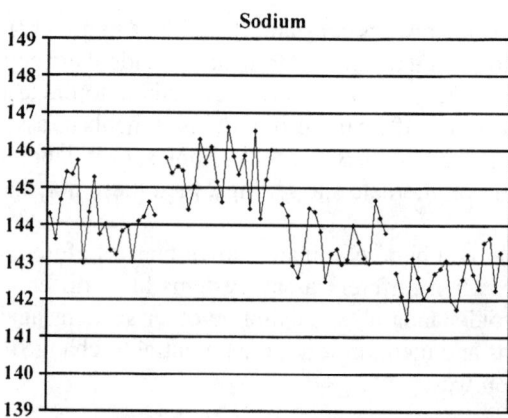

Figure 1.5 Results of Replicate Analyses of One Sample (Sodium, mmol/L) over Time

THE NATURE OF BIOLOGICAL VARIATION

BIOLOGICAL SOURCES OF VARIATION

Some analytes vary over an individual's lifespan, and that is why we often stratify or partition reference values according to chronological age. In addition, certain analytes have predictable biological cycles or rhythms. Thus, if samples from an individual are taken

- at different times of the day when daily rhythms exist,
- at different times during the month when monthly cycles exist, or
- at different seasons (when this is an influencing factor),

then this inherent cyclical variation will contribute to differences in numerical values of results (measured serially) from the same individual.

For many analytes, this is not a major problem because they do not have marked clinically important cyclical rhythms (although one can observe small rhythms in many analytes if very detailed studies are performed). It is vital to know of these rhythms, however. Significant benefits accrue to those who know how to generate or find, and then routinely apply, data on random biological variation.

Understanding Biological Variation In everyday practice, we can use a very simple model to understand intrinsic biological variation—i.e., the random fluctuation around a homeostatic setting point. This random variation is termed the within-subject or intra-individual biological variation.

If we performed the same test on a variety of individuals, we would find that the homeostatic setting points of our subjects would vary. The difference between the homeostatic setting points of individuals is termed the between-subject or inter-individual biological variation.

Certain analytes vary in non-random ways as well. Knowledge of non-random variation is important not only for the correct uses of reference values but also for the correct requesting of tests and for the logical interpretation of results.

Biological Variation during the Lifespan Many analytes do change with age (see Figure 1.6 for examples). Marked changes can occur particularly during critical periods of life:

- neonatal, when there is adaptation to extra-uterine life,
- childhood,
- puberty,
- adulthood (and, for women, the critical period of the menopause), and
- old age.

Many data on age-related changes in analytes are available, and it is impossible to document them all here. Comprehensive compendia of information on reference intervals in children and in the elderly have been published in addition to many general books on reference values in laboratory medicine.

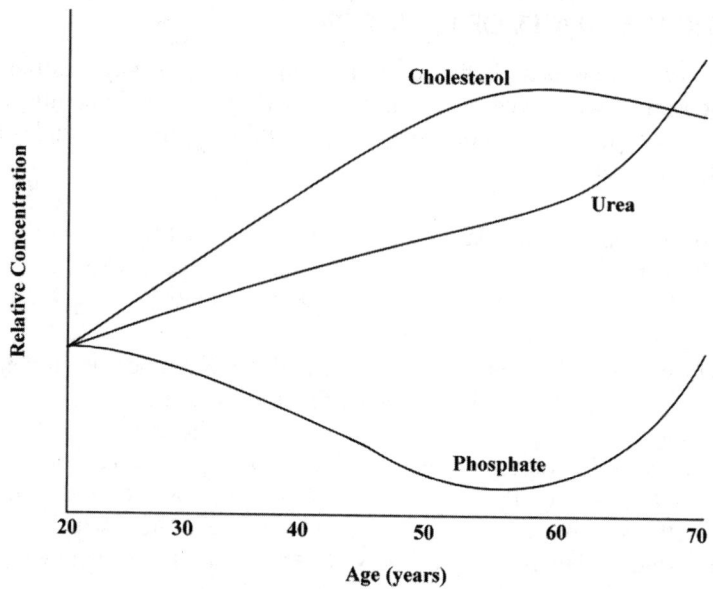

Figure 1.6 Concentrations of Three Serum Analytes in Males Aged 20–70 Years (Analyte Specific Patterns of Age Related Changes)

In practical terms, every test result an individual receives should be accompanied by reference values that are appropriate for the age of the individual on whom the tests have been requested. A real difficulty, of course, is that while we usually know chronological age with great accuracy, it is more difficult sometimes to estimate the all-important biological age.

Daily Biological Rhythms We commonly think of daily biological rhythms as "circadian," but this terminology applies only when the variation depends on the time of day. Some of the so-called daily rhythms that have been documented in the literature actually have little to do with either clock time or sleep/wake (nycthemeral) patterns. Instead, these rhythms are simply due to factors such as changes in posture (proteins and protein-bound constituents fall on changing to a supine position, the usual sleeping position); intake of food (which affects glucose and triglycerides, for example); and intense physical activity (which raises creatine kinase activity).

Because cortisol and growth hormone (and other analytes with daily rhythms) have different cyclical patterns, each analyte must be considered individually. For example, changes to the sleep/wake cycle have an immediate effect on the growth hormone cycle but, in contrast, a delayed effect (several days) on cortisol. In addition, these daily rhythms do not necessarily remain constant over the lifespan. For example, growth hormone concentrations after sleep are usually high just before and during puberty but peaks may be absent in middle to old age.

THE NATURE OF BIOLOGICAL VARIATION

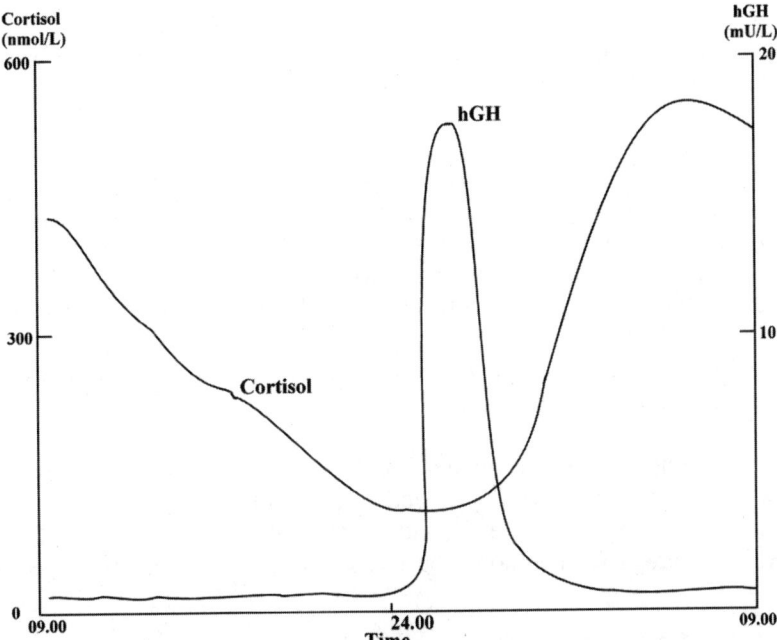

Figure 1.7 Concentrations of Serum Cortisol and Growth Hormone over a Typical Day with a Usual Sleep/Wake pattern

Remember that these figures, like many diagrams in "educational" materials, are smoothed. Both analytes have short spikes or pulses imposed on the basic rhythm.

Those concerned about the effect of biological rhythms on quality in laboratory medicine should consider the following.

- It is impossible to develop good reference values for each point in a given cycle. Generating these values is a complex and time-consuming process (see Chapter 4).
- Ideally, all test samples should be collected in the morning to minimize the influences of daily rhythms (and to reduce pre-analytical variation). At least in hospital practice, laboratory-managed phlebotomy may undertake sampling procedures this way for in-patients.
- Samples for certain analytes, such as serum cortisol, should be taken at important times (for instance, 0900 hours and 2400 hours), when good reference values are available and when the concentration should be at the maximum and minimum.
- The absence of expected rhythms may give important clues as to the presence of disease. For example, failure to show a nycthemeral rhythm in cortisol is characteristic of Cushing's syndrome (over-secretion of cortisol); sampling growth hormone after the patient suspected of deficiency falls asleep is actu-

ally a good clinical investigation to assess whether the expected large rise occurs. However, the preparation of the patient and the requirement to take samples using an indwelling canula means it is rarely performed now.

Monthly Cycles The monthly cycles of women who are in the reproductive phase of their life are of significant importance in laboratory medicine. Like analytes subject to daily rhythms, analytes associated with reproductive cycles present quite different cyclical patterns. Therefore, they must all be considered individually. In addition, a number of other hormones, and even some analytes such as the breast tumor marker CA-153, have monthly cycles, although these are less clinically significant.

The following are important considerations for interpreting test results of analytes affected by monthly cycles.

- It is impossible to develop good reference values for each point during the cycle—their generation is complex and time consuming.
- Samples for certain analytes should be taken at specific, relevant times. For example, progesterone should be measured at day 21 in the cycle if one

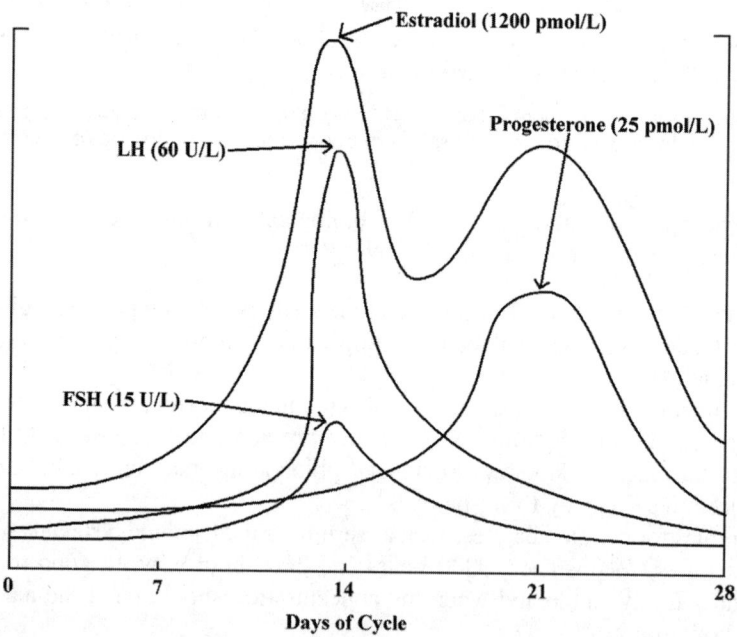

Figure 1.8 Concentrations of Four Hormones During a Classical 28-Day Menstrual Cycle

This figure shows the daily rhythms of four analytes with important monthly cycles: luteinizing hormone (LH); follicle stimulating hormone (FSH); estradiol; and progesterone. Again, note that these are drawn as smooth, rather idealized curves.

THE NATURE OF BIOLOGICAL VARIATION

wishes to assess whether ovulation has occurred and whether the corpus luteum has formed.
- Knowledge of expected cyclical values is vital when these analytes are used either to investigate infertility or in assisted conception programs.

Seasonal Rhythms   Seasonal rhythms have been less well documented than other biological cycles, partly because such studies should be conducted over more than one cycle. Thus, good examinations of seasonal rhythms take years to accomplish. One example is shown in Figure 1.9.

Studying true seasonal variations is quite difficult; many pre-analytical factors influence test results and these may be the main cause of what appear at first to be seasonal effects. For example:

- Serum lactate dehydrogenase activity is higher in summer than in winter. This has been attributed to increased physical activity.
- Serum cholesterol concentration is higher in the winter than in the summer; lower physical activity, increased food intake, and lower levels of sunlight in winter have all been suggested as causes.
- Blood volume tends to increase in summer, due to higher temperatures.
- Proteins increase by about 10% in winter.
- Glycated hemoglobin rises in summer.

These generally small effects, while of academic interest, do not seem to significantly impact on the clinical interpretation of laboratory test results in everyday practice.

RANDOM BIOLOGICAL VARIATION

Although variation can be the result of changes over the span of life and rhythmical biological cycles, the biological variation of many analytes can be described, in the simplest model, as random fluctuation around a homeostatic setting point. In this section, we will explore this model further.

A Case: Test Results over Time in One Individual   Earlier in this chapter, we reviewed two sets of test results for one individual who was apparently healthy. We saw that all results except the serum bilirubin concentration lay within the appropriate age and sex matched reference intervals, and we found the reason for the one "unusual" result.

This particular individual has had a variety of tests obtained over the years at regular intervals. A selection of the data taken over the period 1996–1999 is shown in Table 1.4.

Note that these laboratory results do vary over time and the numbers we obtain are not constant: they appear to vary as random fluctuation around a homeostatic setting point, possibly due to

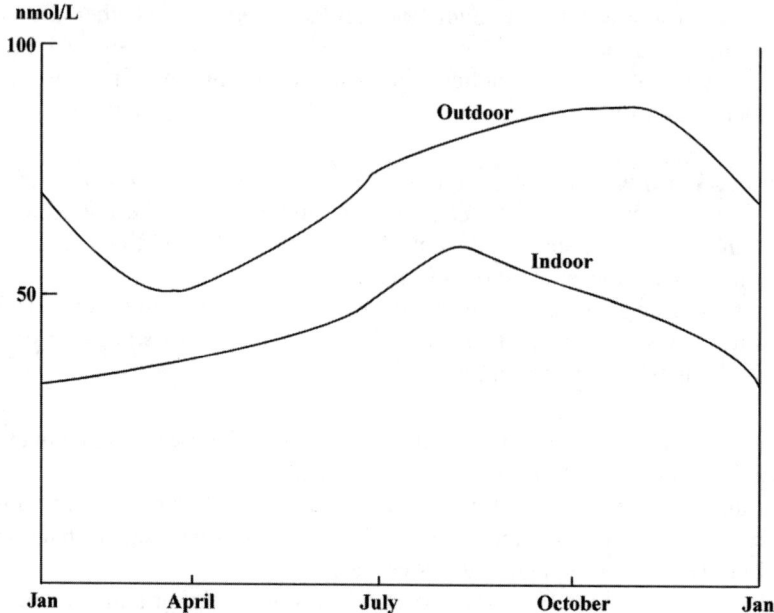

Figure 1.9 Serum 25-Hydroxycholecalciferol Concentrations in Indoor and Outdoor Workers in Dundee, Scotland, (56° North)

A major rhythm, especially in those who live at latitudes far from the equator, is the effect of season on calcium metabolism. In a typical year, maximum ultraviolet radiation occurs from May to July. Concentrations of 25-dihydroxycholecalciferol over one year are shown in this figure. Note that there is a time lag between the environmental ultraviolet radiation peak and the peak value in serum concentrations. In addition, sunlight causes increased cholecalciferol production in the skin, and its metabolites influence calcium homeostasis, increasing serum calcium and increasing urinary calcium output in the summer. (Adapted from Devgun MS, et al. Vitamin D nutrition in relation to season and occupation. Am J Clin Nutr 1981;34:1501–4, with permission).

- pre-analytical factors,
- analytical random variation (precision)—and possibly systematic error (changes in bias), and
- inherent biological variation.

Irrespective of the source of variation, visual inspection of the test results shows that:

- The results for all analytes vary over time.
- Some tests, for example sodium, vary little (137–140 mmol/L).
- Other tests vary more, for example PSA (1.5–2.5 µg/L).
- The mean of certain tests appears close the mean of the reference interval (for example, potassium).
- The mean of certain tests appears to be near the lower reference limit (for example, sodium).

THE NATURE OF BIOLOGICAL VARIATION

Table 1.4 Test Results for One Individual, 1996-1999

Analyte	Units	1st Result	2nd Result	3rd Result	4th Result	5th Result
Sodium	mmol/L	139	139	137	140	138
Potassium	mmol/L	4.3	4.1	4.1	4.4	4.4
Urea	mmol/L	4.0	4.4	4.1	3.9	3.6
Creatinine	µmol/L	88	97	89	82	88
ALT	U/L	40	28	32	33	31
Bilirubin	µmol/L	19	21	17	18	17
Alk Phos	U/L	49	46	52	46	45
Calcium	mmol/L	2.39	2.33	2.25	2.36	2.29
Albumin	g/L	45	48	47	46	47
Cholesterol	mmol/L	4.60	4.82	4.84	4.64	4.41
Triglycerides	mmol/L	0.48	0.52	0.39	0.35	0.43
TSH	mU/L	2.03	2.19	1.89	1.93	2.06
PSA	µg/L	1.5	2.5	2.1	1.8	1.9

- The mean of certain tests appears to be near the upper reference limit (for example, albumin),
- The "unusual" test, bilirubin, is always at or above the upper reference limit, and
- No test varies throughout the entire dispersion of the reference interval.

We now must ask whether these findings are unique to this individual or whether they are a more general phenomenon.

Exploring Random Biological Variation: Serum Creatinine in Men and Women In order to explore random biological variation further, let us examine serum creatinine in men and women. Serum creatinine was measured four times in a group of men and a group of women at intervals of 14 days. The results are reported in Tables 1.5 and 1.6 and shown in Figures 1.10 and 1.11.

Visual inspection of the numerical test results and the bars in both figures shows that

- All individuals change over time.
- No individual has values that span the entire reference interval.
- The range of values from one individual occupies only a small part of the dispersion of the reference interval.
- Most individuals' values lie within the reference interval.
- The mean values of all individuals lie within the reference interval and differ from each other.
- Most individuals could have values that were very unusual for them but these values could still lie well within the reference interval.
- Individuals can have values that span the lower and upper reference limits respectively, and thus, individuals can have values that range from usual to unusual over time (or *vice versa*).

Table 1.5 Serum Creatinine Concentrations (μmol/L) in 10 Apparently Healthy Men

Man	1st Result	2nd Result	3rd Result	4th Result
1	60	63	66	62
2	103	99	110	107
3	88	85	93	86
4	125	120	115	118
5	75	83	78	86
6	92	98	90	96
7	75	70	68	71
8	105	110	99	103
9	72	81	74	78
10	68	75	72	77

Interpretation of Creatinine Concentration Results These data show that as far as serum creatinine is concerned, each individual has his or her own individual homeostatic setting point, and that each individual's results span only a small part of the population-based reference interval. (This is referred to as the high individuality of creatinine. As we will see later, most analytes have this marked individuality.) These variations around setting points can be due to pre-analytical variation, analytical variation (precision and changes in bias) and within-subject biological variation.

The difference between homeostatic setting points themselves is the *between-subject* biological variation. Moreover, in health, the value of the setting point for each individual is probably most dependent on muscle bulk, so it is hardly surprising that men have, in general, higher homeostatic setting points than women.

It is important to recognize that individuals can have values that are highly unusual for them—and that these values still lie within the population-based refer-

Table 1.6 Serum Creatinine Concentrations (μmol/L) in 10 Apparently Healthy Women

Woman	1st Result	2nd Result	3rd Result	4th Result
1	61	64	66	59
2	45	50	52	55
3	79	72	74	78
4	83	77	86	79
5	95	104	99	97
6	65	68	69	63
7	92	86	94	89
8	77	73	78	71
9	65	68	75	71
10	89	83	85	82

THE NATURE OF BIOLOGICAL VARIATION

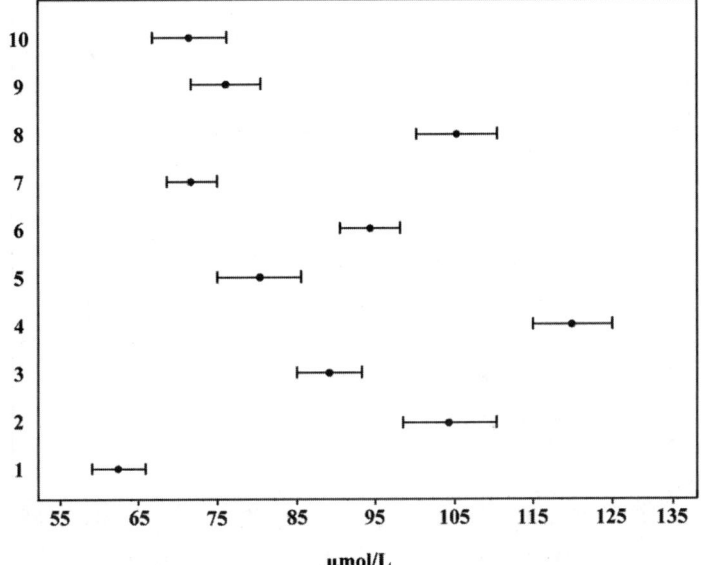

Figure 1.10 Mean Values and Absolute Ranges of Serum Creatinine in Four Samples Taken from Each of 10 Apparently Healthy Men.

The age and sex matched reference interval for men aged 18–55 years is 64–120 μmol/L.

ence interval—even if these values are stratified or partitioned according to age and/or sex. This simple fact provides logical explanations why, in general:

- Laboratory tests are not very good at picking up early or latent disease.
- Screening using routine laboratory procedures is not very productive.
- Case finding using common laboratory tests has not been as useful as was hoped when multi-channel analyzers appeared 30 years ago.
- Given that some individuals vary quite naturally above and below reference limits, tests that are "slightly abnormal" are often "normal on repeat." (We will explore this in greater detail in Chapter 4.)

By carefully controlling pre-analytical variation, and by designing experiments so that we can quantify analytical variation, we can obtain pure estimates of the average within-subject biological variation and the between-subject biological variation, which we can use

- to set quality specifications,
- to assess the significance of changes in serial results in an individual,
- to consider the utility of conventional population-based reference values, and
- in other applications.

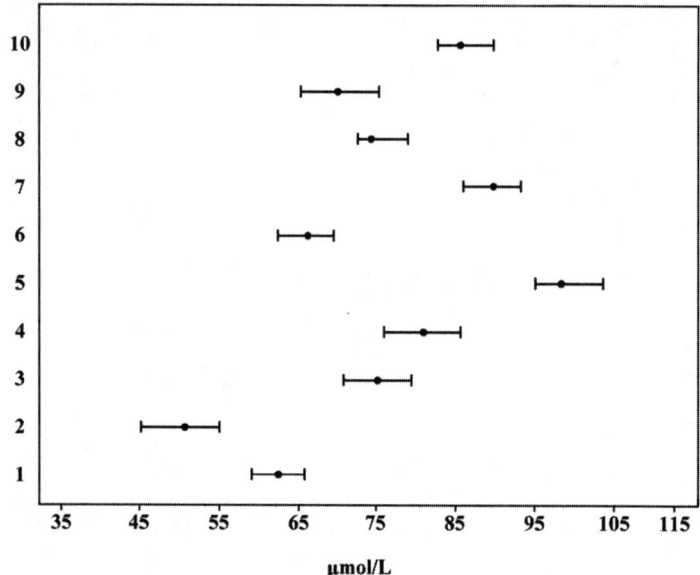

Figure 1.11 Mean Values and Absolute Ranges of Serum Creatinine in Four Samples Taken from Each of 10 Apparently Healthy Women

The age and sex matched reference interval for women aged 18–55 years is 50–100 μmol/L (note that this is not the same as the reference interval for men, even though the groups are in the same age range).

DETERMINING COMPONENTS OF BIOLOGICAL VARIATION

Before we can apply data on within- and between-subject components of biological variation, we must explore how to determine these quantitatively. For example, it is current dogma that all laboratories should generate their own population-based reference values for every test in their repertoire. By analogy, then, is it necessary for all laboratories to determine their own data on biological variation? Moreover, if in fact we can use existing data in our everyday medical laboratory practice, we must investigate sources of good data on biological variation.

Selection of Subjects for Studies Generating data on the components of random biological variation, i.e., within-subject and between-subject biological variation, is in some ways like generating conventional population-based reference values. The difference is the number of subjects studied, and the number of samples taken from each: instead of taking one sample from a large group of people, we take a number of samples from a smaller group of subjects. Since we are mostly interested in the biological rather than any pathological sources of variation, we usually investigate healthy people, although there are some data derived from groups with specific diseases in the literature.

The subjects selected for study should be "reference individuals." The usual

THE NATURE OF BIOLOGICAL VARIATION

approach is to use what the International Federation of Clinical Chemistry (IFCC) has termed the *a priori* approach—which simply means that we adopt some exclusion (or inclusion) criteria *before* the subjects are involved in the experiment.

Usually, pre-analytical variation is minimized as far as practicable. In part, this means that we pick people who are

- willing to provide a number of samples over a period of time,
- apparently healthy, and not taking any drugs that might affect the analytes under investigation (including contraceptives and over-the-counter medications),
- not practicing unusual lifestyles or habits, and
- not consuming more than the recommended number of units of alcohol (use of tobacco products should probably also be an exclusion criterion).

Some have suggested that we should perform a battery of laboratory tests to check for health before recruiting individuals into such studies. Clinical, biochemical, and hematological tests might be done before subjects are accepted. However, if all the suggested criteria were rigidly applied, we would have only a few people left to study. Usually we like to recruit both men and women into studies on biological variation since gender is often an important factor in the interpretation of laboratory test results.

How many subjects do we need, then, to perform a study on biological variation? There is no definite answer. The more subjects we have, the better the estimates will be. In statistical terms, the confidence intervals will be smaller if we have more data. However, the more subjects we have, the more difficult it will be to collect the samples, handle these appropriately, and do the analyses under the right conditions. Thus, the number of subjects is a compromise between the large number that is the ideal and the smaller number that can be handled in any good experimental design.

Sample Collection, Handling, and Storage Do everything possible to minimize pre-analytical variation in subjects—we need the best estimates of *biological* components of variation. This includes

- taking samples at the same time of day—usually early morning—unless the experiment is intended to look at biological variation over periods of less than one day,
- taking samples under the same conditions: no strenuous exercise before sampling, a standard meal or, perhaps better, no breakfast, and sitting down for at least 30 minutes prior to sample collection,
- taking blood samples with a standard phlebotomy technique, preferably with a single experienced phlebotomist, into collection tubes of the same lot number,
- transporting samples to the laboratory under identical conditions of temperature and elapsed time, and
- centrifuging, when required, at the same speed and temperature for the same period of time.

We can take two possible approaches, then: (1) freeze all the samples (this is the ideal option provided that analytes are stable under the storage conditions); or (2) analyze the samples as they are collected (required if the analyte is unstable).

If fluids other than blood are the subject of study, then follow similar principles. For example, to study the biological variation of analytes in 24-hour urine samples, then:

- Give identical explicit and unambiguous instructions to subjects.
- Standardize start times and stop times.
- Keep the amount of stabilizer or preservative in the collection container constant.
- Have only one person weigh or measure volume with one set of equipment.
- Ideally, store aliquots of samples until they can be analyzed in conditions that ensure stability.

Analysis Keep analytical variation as low as possible; in fact, adopt techniques that give what has been termed "optimal conditions precision."

To have good precision, and negligible bias, the analyses ought to be done, ideally, with

- one instrument,
- one operator,
- one set of calibrators,
- one lot of reagents, and
- single lots of any ancillary reagents and consumables.

The best experimental design assays samples twice in random duplicate in a single analytical run. This has real advantages.

- Between-run analytical variation is eliminated.
- The analytical component of variation is derived from replicate analyses of subject samples, ensuring that analytical variation is estimated at the same level as subject samples, and that the matrix is identical throughout.

Those who have worked extensively in this field favor the approach above. There is one problem, however: the intensive analytical effort required to analyze samples in this manner limits the number of subjects and samples that can be studied.

Another relatively common protocol is to collect and store all samples as described above and then analyze them only once in one analytical run. In this approach, analytical variation must be assessed using quality control samples. This has some disadvantages: for some tests, precision achieved with quality control samples differs from that attained with patient samples. Thus, we really should *prove* that the precision of the analyses is the same for both types of sample. We can do this simply by running some of each in replicate, and calculating the precision as SD (or CV) from the replicates. This is done using the formula

SD = the square root of (the sum of the squares of differences between each of the pairs divided by twice the number of pairs)

that is,

$$SD = (\Sigma d^2 / 2n)^{1/2}$$

Then we compare the SD (or derived CV), using the simple F-test, which involves calculating the ratio of the variances (the SD^2) to generate the F value and comparing the value with the critical values of F in standard statistical tables. Of course, we should use control materials with analyte levels of the same order as the values for the analyte in subject samples.

If samples are unstable or have been analyzed as they were collected, then we must estimate between-run precision from quality control materials. This might create problems. If the samples were assayed in duplicate on the day of collection, we could easily calculate within-run precision from the duplicates as shown above. However, we would have to estimate within-run and between-run precision from analyses of quality control materials. If the within-run precision of samples and quality control materials did not significantly differ (using the objective F-test), we could assume that the between-run precision was also comparable. Thus, simply by subtraction, we could obtain good estimates of the components of biological variation. As we will see later, it is fairly easy to add and subtract sources of random variation, provided we use the variances.

Initial Inspection and Statistical Treatment of Raw Data We can describe the statistical analysis of results rather simply. However, the actual calculations are relatively complex when done without the aid of statistical software. The methodology has been described in great detail by Fraser and Harris with a very full description of a numerical example. Given the experiment and the ideal approach as described above, we have

- duplicate data on each sample,
- series of duplicate results on samples taken over time from every individual in the cohort of subjects studied, and
- the set of duplicate results obtained on the group.

First, we examine the data at these three levels for outliers—i.e., numbers that differ from the bulk of the data set. We need to do this because even a single unusual observation, perhaps resulting from an analytical mistake or simple misidentification of a sample, can greatly influence the estimates of the components of variation.

Statistically, we can use two tests to look for outliers: (1) the Cochran test and

(2) Reed's criterion. It is probably useful to seek the aid of a medical statistician to apply these. The principles are as follows.

- We can look for outliers in the sets of duplicate results using the Cochran test that examines the ratio of the maximum variance to the sum of the variances and compares this to the appropriate critical values in statistical tables. If we find one outlying variance (an unexpected large difference in the duplicates), we reject both data points.
- We can again use the Cochran test to look for outliers in the variances of the results from each subject to see if any individual's dispersion of results is larger or smaller than that of the group as a whole. This is called *examining the heterogeneity of within-subject biological variation*.
- We can then see if any individual has a mean value that differs greatly from the other subjects, using Reed's criterion. This very simple and widely used statistical test considers the difference between the extreme value and the next lowest (or highest) value and rejects the extreme value if this difference is more than one-third of the absolute range of values (highest minus lowest).

One of the best ways to look for outliers is simply to write down all the duplicate results and look at the numbers to see if there are any obvious differences between duplicates. Then, create a simple diagram of the mean values and the absolute ranges of values, for evaluation as shown in Figure 1.12.

Estimating Components of Biological Variation   Once we have detected and eliminated outliers, the numbers we now have (from the ideal experiment for further statistical analysis) has an observed variability that is made up of

- the average variance of the duplicate assays (within-run precision),
- the average within-subject biological variance (the variance around the homeostatic setting point from time to time in the average subject), and
- the variance of the true means (the homeostatic setting points) among subjects, i.e., the between-subject variance.

These variances are best derived using nested analysis of variance (ANOVA), available in statistical software packages. (The calculations can be done manually, but this is rather complex and time-consuming. The review by Fraser and Harris presents an example as a model. Note that this is not simply a matter of addition of variances, although this (somewhat naïve) approach has been used many times in the literature on biological variation.)

If we do not use the ideal experimental approach, then we usually calculate the between-subject variation and the other residual component (due to both analytical and within-subject biological variation). Then we can subtract the analytical component, usually in this case by simple calculation, subtracting variances—since total SD^2 or CV^2 is simply the sum of the component SD^2 or CV^2. We will explore this concept in detail later in this book.

THE NATURE OF BIOLOGICAL VARIATION

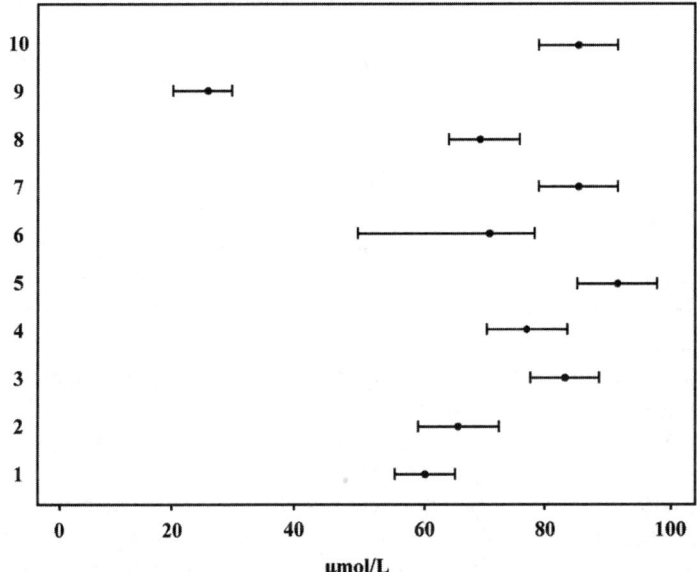

Figure 1.12 Graph of Data Showing Means and Absolute Ranges for a Hypothetical Not-Yet-Studied Analyte

After examining the list of duplicate results to scan for outliers, visually inspect a graph of the data. The graph shown here demonstrates that Subject 6 has a larger within-subject plus analytical variation range than the other individuals studied. Moreover, the dispersion does not appear symmetrical. This subject may be an outlier and, formally, the Cochran test would be applied to assess whether the variance for this subject was indeed statistically different from the others. In addition, the mean value for Subject 9 seems to be lower than the mean values for other subjects, and Reed's criterion would be applied to test whether all values for this subject should be excluded.

This hypothetical example of an investigation of the biological variation of a new analyte shows that, when represented graphically, (1) any mean value that is really different from the others (an outlier) becomes abundantly clear; and (2) results from any individual whose range of values differs greatly from the other subjects (again—an outlier) stand out.

Today it is all too easy to type a set of data into a statistical package and perform many types of statistical manipulation. Inspecting the data before doing the calculations is absolutely vital. This is a very important concept that is often ignored now with the ubiquitous availability of personal computers and statistical packages.

Irrespective of how we perform the calculations (or have them performed for us) and only after we have decided the quality of the data, we have numerical estimates of

- analytical precision—as SD or CV—usually denoted as SD_A or CV_A (in the ideal experiment, this will be an estimate of the optimal conditions within-run precision),

- average within-subject biological variation (as SD or CV) usually denoted as SD_I or CV_I, and
- between-subject variation (as SD or CV), usually denoted as SD_G or CV_G.

The Constancy of Within-Subject Biological Variation Generating numerical data on biological variation is time consuming and clearly requires significant analytical and statistical expertise. Must every laboratory generate its own data on components of biological variation?

There have been many studies on biological variation of commonly requested analytes. Some studies on the within-subject biological variation of serum sodium and urea are shown in Table 1.7. In general, estimates of within-subject biological variation in particular are constant, irrespective of the number of subjects, the time scale of the study, the methodology, and the country in which the study was done.

It has been suggested that homeostatic mechanisms may diminish during the aging process, that biological variation increases with age, and that biological variation is smallest between 30 and 50 years of age. We have studied the biological variation of commonly requested biochemical and hematological analytes in older people. The biological variation of a few commonly requested serum analytes in younger people and healthy people over 70 years of age is shown in Table 1.8. There is little evidence that biological variation differs in younger and older individuals. Moreover, this hypothesis that the within-subject biological variation is constant does not only hold true for serum analytes. Studies on within-subject biological variation of some analytes in 24-hour urine collections are shown in Table 1.9.

This constancy is hardly surprising because within-subject biological variation is the quantitative estimate of homeostasis in a single animal species—humans!

Existing Databases We do not need to generate data on components of biological variation in every laboratory because there is ample evidence that estimates of biological variation are constant. Indeed, in ill people, if the analyte is unaffected

Table 1.7 Some Studies on the Within-Subject Biological Variation of Serum Sodium and Serum Urea (CV, %)

No. of subjects	Time span (weeks)	Sex	Sodium	Urea	Country
11	2	M	0.7	12.3	Denmark
10	4	M	0.9	14.3	USA
10	8	M	0.6	9.5	Germany
14	8	F	0.5	11.3	Germany
9	12	M	1.4	13.6	USA
11	15	M	0.6	15.7	Denmark
37	22	M	0.5	11.1	England
15	40	M&F	0.7	13.9	Scotland

Table 1.8 Some Studies on Within-Subject Biological Variation in Younger and Older Individuals in Dundee (CV, %)

Analyte	Younger	Older
Sodium	0.7	0.9
Potassium	5.4	4.8
Chloride	1.2	1.2
Urea	13.9	10.3
Creatinine	4.1	4.3
Calcium	2.1	1.6
Cholesterol	4.9	5.8
Proteins	3.1	2.6
Albumin	2.2	2.6

by the disease, the biological variation is of the same order as in healthy individuals. Even if the analyte is actually affected by the disease, if the illness is chronic and the disease state is stable, again there is evidence that within-subject biological variation is of the same magnitude as in healthy people when expressed in CV. Homeostatic setting points have changed, probably due to pathology, but variations around these have not.

In consequence, the existing databases on biological variation can be used in all laboratories. In 1982, John Ross, in an essay on evaluation of precision, compiled published data that listed the study's time span, the CV_I and CV_G, other indices, and references. Callum Fraser followed this type of model in 1988 and then updated this work in 1992. The references to these rather dated compilations can be found in the comprehensive database published by Xavier Fuentes-Arderiu and his colleagues in 1997. This database (available at www.westgard.com/intra-inter.htm) summarizes median CV_I and CV_G.

The most recent and extensive compilation of data on components of biological variation and many useful derived indices has been provided by Carmen Ricos and her associates from Spain (available at www.westgard.com/guest17.htm). It is

Table 1.9 Some Studies on Within-Subject Biological Variation of Urine Output (CV, %)

Analyte	Men: Australia	Men and women: Scotland	Men and women: Spain
Sodium	28.0	26.5	28.7
Calcium	25.1	26.2	27.5
Creatinine	11.2	11.0	15.0
Phosphate	16.6	16.9	20.6

highly recommended that laboratories everywhere use this source of information. Appendix 1 of this book contains three tables with data on within-subject (CV_I) and between-subject (CV_G) components of biological variation for commonly assayed analytes in three sections, namely

- serum or whole blood (indicated in parentheses),
- urine, and
- hematology and hemostasis.

Additional data on very rarely measured analytes do exist. Before determining data on analytes not included in these three tables, readers are urged to search the original databases to assess whether data already exist.

SUMMARY

This chapter has covered the following points.

- Test results obtained on samples from individuals vary over time.
- There are many sources of pre-analytical variation in the preparation of subjects for sample collection.
- Sample collection procedures have many possible sources of variation.
- Random analytical variation (precision) contributes to test result variation.
- Systematic variation (bias), particularly changes in bias over time, also contributes to test result variation and should be minimized.
- Biological variation is a source, often the most important, of variation in results over time.
- Some analytes vary through the lifespan, particularly at times of rapid physiological change.
- Some analytes, particularly hormones, have marked predictable cyclical rhythms, which are relevant to the generation of reference values, collection of samples at suitable times, and in the diagnosis of disease.
- Most analytes vary randomly around a homeostatic setting point. This is termed within-subject biological variation and the usual abbreviation is CV_I.
- For most analytes, different individuals have different setting points, and this difference is termed the between-subject biological variation. The usual abbreviation is CV_G.
- Components of biological variation can be generated by duplicate analysis, under optimal conditions, of a set of samples carefully collected from each of a cohort of reference subjects under standardized conditions. After excluding outliers, nested analysis of variance is the preferred statistical approach.
- There are many data on biological variation in the literature. All laboratories should use available data to set quality specifications, derive reference change values and delta-check values, assess the utility of conventional reference values, and for other purposes.

FURTHER READING

1. Brooks, Z. Performance-driven quality control. Washington, DC: AACC Press, 2001.
2. Fraser CG, Harris EK. Generation and application of data on biological variation in clinical chemistry. Crit Rev Clin Lab Sci 1989;27:409–437.
3. Fraser CG. Interpretation of clinical chemistry laboratory data. Oxford: Blackwell Scientific, 1986.
4. Guder WG, Narayanan S, Wisser H, Zatwa B. Samples: from the patient to the laboratory. Darmstadt: GIT Verlag GmbH, 1996.
5. Ricos C, Alvarez V, Cava F, Garcia-Lario JV, Hernandez A, Jimenez CV, Minchinela J, Perich C, Simon M. Current databases on biologic variation: pros, cons and progress. Scand J Clin Lab Invest 1999;59:491–500. Available online: www.westgard.com/guest17.htm.
6. Sebastian-Gambaro MA, Liron-Hernandez FJ, Fuentes-Arderiu X. Intra- and inter-individual biological variability data bank. Eur J Clin Chem Clin Biochem 1997;35:845–852. Available online: www.westgard.com/intra-inter.htm.
7. Young DS. Biological variability. In: Brown SS, Mitchell FL, Young DS, eds. Chemical diagnosis of disease. Amsterdam, New York, Oxford: Elsevier North Holland Biomedical Press.1979:1–113.
8. Young DS. Effects of pre-analytical variables on clinical laboratory tests. Washington, DC: AACC Press, 1993.

Chapter 2
Quality Specifications

Modern quality management involves much more than the simple statistical quality control performed every day at the bench. The essential elements of quality laboratory practice, quality assurance, quality improvement, and quality planning must be included in quality management. Indeed, these elements constitute the basic elements of total quality management in the laboratory medicine setting.

All definitions of quality, and there are many, can be interpreted in our field as *establishing conditions such that the quality of all tests performed in laboratory medicine assists clinicians in practicing good medicine*. Therefore, before we can control, practice, assure, or improve laboratory quality, we must know exactly what level of quality is needed to make sure that satisfactory clinical decision-making is possible. Thus, specifying the quality required is a necessary prerequisite for instituting quality management (see Figure 2.1).

SETTING QUALITY SPECIFICATIONS

The level of performance required to facilitate clinical decision-making has been given a number of names. The currently most widely applied term is *quality specifications*. Other terms include *quality goals, quality standards, desirable standards, analytical goals,* and *analytical performance goals*.

If you asked various people concerned with the production of test results and others involved in requesting tests to define a good laboratory test, each person would likely give a very different answer. For example:

- The laboratory director might reply, "A test that always performs well in PT and EQAS and allows us to maintain our accredited status."
- The laboratory manager might answer, "A test that is inexpensive and easy to do with relatively unskilled and cheap to hire staff."
- The technologist might respond, "A test that never fails the internal quality control."
- The emergency room clinician might reply, "A test that can be done very rapidly at the bedside and on whole blood."
- The academic physician might answer, "A test with high clinical sensitivity, specificity, and predictive value."
- The pediatrician might respond, "A test that requires a very small sample volume."

These hypothetical answers reflect the fact that every laboratory test has many different attributes, best termed *performance characteristics*. Every method can

Quality Management

Figure 2.1 The Central Role of Quality Specifications in Quality Management

be fully described in terms of its performance characteristics, which fall into two classes.

- *Practicability characteristics* are concerned with the details of executing the procedure, and include many facets such as skills required, speed of analysis, volume of sample required, and type of sample able to be analyzed.
- *Reliability characteristics* deal with the scientific facets of the method such as precision, bias, limit of detection, and measuring range.

Ideally, quality specifications should be available for each and every performance characteristic of laboratory procedures, particularly for reliability characteristics, and especially for precision and bias. To implement a proper laboratory quality management system, we must be able to define quality specifications for precision and bias, and then for total error allowable.

USES OF QUALITY SPECIFICATIONS

The need for objective quality specifications in many aspects of laboratory quality management can be very well illustrated by thinking about how any new analytical system, instrument, or method is introduced into service in the clinical laboratory.

These steps include

- documenting requirements,
- assessing available systems,
- preparing a specification,
- creating a short list for evaluation,
- carrying out method evaluation or validation and assessing evaluation data,
- instituting a planned internal quality control system, and
- enrolling in an appropriate PT or EQAS.

QUALITY SPECIFICATIONS

Documenting Requirements It is essential to define objective quality specifications at the outset of the process. This first step in the introduction of any new technology must be done carefully and with considerable thought. We must document in detail the desirable performance characteristics with respect to both practicability and reliability. We must define what we want as test repertoire; sample matrix (serum, plasma, CSF, urine, fluids); sample volume (adults, children, neonates); time for STAT and routine tests and throughput; chemistry of the methodology; reagent pack size; assignation of values to calibrants; calibration frequency and stability; number of quality control samples on board; and spectrum of quality control rules on board. We should delineate the space we have available (the footprint) and which services (power, water, light, cabling) are already on hand and possible. We should know what funds we have available now and in the future. And, most importantly at this stage, we should define the quality of precision, bias, and total error we want, and, in addition, the limit of detection, measuring range, interference, specificity, and carry-over.

Assessing Available Systems Once we have defined exactly what we need, we can assess what is available to potentially meet our needs. We can consult peer-reviewed journals, the trade press, and articles in the house journals of major manufacturers. We can study manufacturers' advertisements and data, and attend their seminars or workshops, particularly at conferences and congresses. We can visit other laboratories and discuss the pros and cons of solutions reached by colleagues. We can study the wealth of information available in reports from PT and EQAS. We can then use our pre-set quality specifications to compare our desired specifications with what is technically and methodologically possible and available.

Writing a Specification When we have assessed what is available, we may go back and modify our definition of what is needed. We then should prepare a detailed list of both essential and desirable characteristics. This can be used in the preparation of a tender document for potential suppliers to bid for our business. The specification and the tender document should contain detailed numerical quality specifications for as many performance characteristics as possible. We should do this at least in part to remind manufacturers that the reliability characteristics of methods do affect clinical decision-making and are still important considerations in laboratories.

Creating a Short List Once manufacturers and suppliers have responded to the specification or tender document, create a short list of possible solutions for the laboratory. Then compare manufacturers' performance claims for each of the reliability characteristics against predetermined quality specifications.

Analyses of Assessment or Evaluation Data Candidate analytical systems or instruments are often assessed briefly or evaluated in detail before purchase or lease, and certainly before introduction to service in the laboratory. There are many excellent published protocols detailing how to perform an evaluation or method validation. These generate significant amounts of data on performance characteristics. These data should be compared to the desired quality specifications in order to make educated judgements on acceptability. (This process will be discussed later in this Chapter.)

Setting Up the Internal Quality Control System   When introducing the analytical systems or instruments into service, set up a good quality control system and simultaneously introduce all other facets of quality management. Quality planning is essential to deciding how many controls to run and which quality control rules to apply for acceptance or rejection, and this cannot be done without the detailed use of quality specifications. (These processes lie outside the scope of this book but are dealt with in detail in many recent publications.)

Enrolling in PT or EQAS   It is usual, and sometimes mandatory, for laboratories to participate in PT or EQAS for as many analytes as possible. The best of these programs and schemes use objectively set quality specifications to generate the fixed limits against which acceptability is judged.

The need for objective quality specifications in method evaluation and quality control is well documented. For example, in 1999, one of the premier journals in laboratory medicine, *Clinical Chemistry*, stated in its instructions to potential authors that "results obtained for the performance characteristics should be compared objectively with well-documented quality specifications: published data on the state of the art, performance required by regulatory bodies such as CLIA '88, or recommendations documented by expert professional groups." Moreover, the National Committee for Clinical Laboratory Standards (NCCLS) recently updated its guidelines on statistical quality control for those working in the U.S. The revised guidelines include information on planning a statistical quality control procedure, and the first requirement is to define quality specifications.

PROBLEMS WITH SETTING QUALITY SPECIFICATIONS

Quality planning has revolutionized internal quality control systems. However, some suggest that it is too difficult to set quality specifications in the quality planning process and prefer to adhere to traditional statistical process control. Others suggest additional objections to using numerical quality specifications, such as the following.

- There are many published recommendations in books, reviews, and papers from all parts of the world, and it is difficult for the non-expert to decide which models are good and which have problems, making it challenging to select the most appropriate quality specification for use in quality planning.
- Test results are used in a great variety of clinical situations including research and development, teaching and training, monitoring, diagnosis, case finding, and screening. There may be no single set of quality specifications that would make any method suitable for all clinical purposes.

QUALITY SPECIFICATIONS

- As time goes on, new recommendations continue to be published, even by the experts who perhaps seem to keep changing their views and recommendations. This might suggest that there is no really ubiquitous professional consensus about the best way to set quality specifications.
- Some have said that there appears to be no proof that patients (or clinicians) have been harmed by the performance levels of current methodology and technology, and they question changing what has served us well for years.
- Where legislation involving regulation through PT rather than educational types of EQAS exists, such as U.S. CLIA '88 requirements, laboratory efforts might be directed mainly to achieving the standards required to pass, so that, by default, the fixed limits laid down in PT become the quality specifications applied in practice.
- Manufacturers of analytical systems for the clinical laboratory do not appear to use objective professionally set specifications as major considerations in either development or marketing but are driven largely by the state of the art and by what is technically achievable at reasonable cost.

Irrespective of these alleged difficulties, quality specifications are the crux of quality planning and quality management. Knowledge about their creation and application is vital to the modern clinical laboratory operation.

THE HIERARCHY OF MODELS FOR SETTING QUALITY SPECIFICATIONS

Much is written about setting quality specifications: original papers, review articles, and general texts on laboratory medicine. Special thematic conferences have been held to discuss the subject. Therefore, one of the arguments against setting them—namely, that there are many published recommendations and it is not easy for the non-expert to decide which models are good and which have problems—does seem to have some merit.

Because of this, a conference was organized in Stockholm in April 1999 and sponsored by the International Union of Pure and Applied Chemistry (IUPAC), the International Federation of Clinical Chemistry and Laboratory Medicine (IFCC), and the World Health Organization (WHO) to discuss whether or not a consensus could be reached on global strategies to set quality specifications in laboratory medicine irrespective of whether the laboratory was large or small, private or public, or developed or developing. Those who had published original work on models for setting quality specifications were invited to address participants from 23 countries.

The conference achieved its aim: the papers and consensus statement have been published in a special issue of the *Scandinavian Journal of Clinical and Laboratory Investigation*. The consensus statement set the available models into a hierarchical structure (see Table 2.1).

The hierarchy was based upon the proposals in an earlier editorial in *Clinical Chemistry*. Models higher in the hierarchy were preferred to models lower in the hierarchy, and it was recommended that appropriate models be used for particular

Table 2.1 Hierarchy of Strategies to Set Quality Specifications

Rank	Strategy	Subclasses
1	Assessment of the effect of analytical performance on specific clinical decision making	Quality specifications in specific clinical situations
2	Assessment of the effect of analytical performance on general clinical decision-making	A. General quality specifications based on biological variation
		B. General quality specifications based on medical opinions
3	Professional recommendations	A. Guidelines from national or international expert groups
		B. Guidelines from expert individuals or institutional groups
4	Quality specifications laid down by regulation or by EQAS organizers	A. Quality specifications laid down by regulation
		B. Quality specifications laid down by EQAS organizers
5	Published data on the state of the art	A. Published data from PT and EQAS
		B. Published individual methodology

clinical purposes. However, these recommendations were not set in stone because new and better models that might become available could then be incorporated in due course after acceptance by professionals.

One of the difficulties in comparing quality specifications advocated in the hierarchy is that specifications are presented in rather different formats. (This is also true for previous publications on the subject.) Some specifications address precision, some bias, and others total error allowable.

Quality specifications for total error allowable set acceptability criteria for the combined effects of random variation and systematic variation. Many would suggest that physicians think in terms of total error; ideas on quality planning require using total error quality specifications; and the fixed limits in vogue in PT and EQAS are also formatted as quality specifications for total error allowable. Thus, it is vital to examine how to calculate total error before we consider the hierarchy for setting quality specifications and the real meaning of the models' outcomes.

THE TOTAL ERROR CONCEPT

Total error (TE) can be calculated in a number of ways. The most usual way is to add bias and precision "linearly." Note that, in these calculations, it is the absolute

QUALITY SPECIFICATIONS

value of the bias that is used. It does not matter whether the bias is positive or negative. Again the literature has a number of recommendations, including

- the sum of bias plus 2 times the precision, or TE = bias + 2SD (or CV),
- the sum of bias plus 3 times the precision, or TE = bias + 3SD (or CV), and
- the sum of bias plus 4 times the precision, or TE = bias + 4SD (or CV).

However, much of the basic literature on the theory and practice of quality planning uses the following formula for total error allowable (TE_a)

- the sum of bias plus 1.65 times the precision, or TE_a = bias + 1.65SD (or CV). (The basis of this formula is shown in Figure 2.2.)

SD is used when working in terms of the units used for reporting the results and CV is used when variations and errors are considered in terms of percentages since, as we know, CV = (SD/mean) * 100.

The formula we are using here for total error allowable is derived as follows. We usually work at 95% probability to allow for 5% error. As shown in the figure, the values we want to exclude are only at one side of the distribution. Thus, we have 5% excluded at both the upper and lower ends of the distribution; 10% in total. We are therefore including only 90% of the distribution and the appropriate multiplier for this is 1.65. These multipliers are known as Z-scores and we will explore their use later.

The formula for total error allowable then becomes

total error allowable = bias + Z * precision, or
total error allowable = bias + 1.65 * precision for 95% probability, or
$$TE_a = B_A + 1.65 CV_A.$$

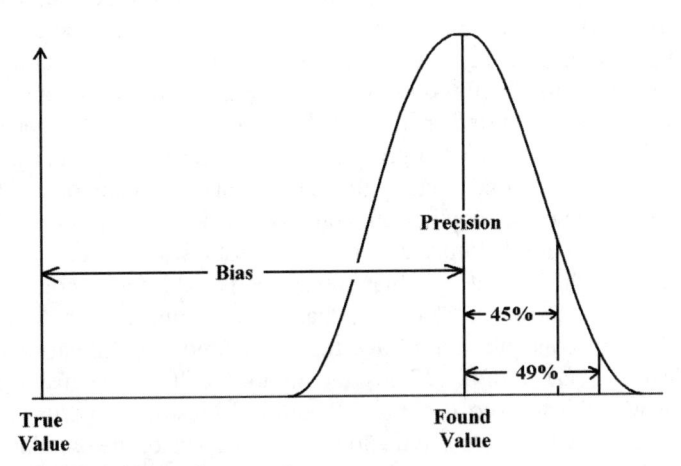

Figure 2.2 The Total Error Concept

STRATEGIES FOR SETTING QUALITY SPECIFICATIONS

Not all strategies for setting quality specifications are included in the hierarchy. Some models found in the literature, especially those in standard textbooks, have many disadvantages and are considered obsolete.

Available models that are regarded by professionals as still having some merit have been set into the hierarchy shown in Table 2.1. However, the inclusion of any particular strategy does not mean that it has no disadvantages. The bases for the models included, and their pros and cons, are the subjects of this section.

Quality Specifications in Specific Clinical Situations Ideally, quality specifications should be derived by numerically assessing the effect of analytical performance on specific clinical decisions. Thus, test by test, and clinical situation by clinical situation, we derive quality specifications directly related to clinical outcomes. It is hardly surprising that this approach is at the top of the hierarchy. Unfortunately, this approach is very difficult and the calculations have been done only for a few analytes in a limited number of different clinical settings.

Here, we will adapt a model examined by Per Hyltoft Petersen and his co-workers in Scandinavia. Let us consider the theoretical example of serum cholesterol when used as a screening test (although this, too, has its advantages and disadvantages) and assume that it has a true population distribution as shown in the top diagram of Figure 2.3. We have assumed that serum cholesterol has a Gaussian distribution and that there is wide agreement on a fixed concentration for clinical action.

Now, if the laboratory's analytical bias is positive, then the curve will move to the right as shown in the middle diagram of Figure 2.3. A greater number of the population would now be above the chosen fixed limit for clinical decision-making, including individuals who really had a high serum cholesterol concentration above the fixed limit and those who had high concentrations simply because of the positive analytical bias. Thus, "false positive" results would be found.

Thus, the performance characteristics of the analysis itself do affect clinical outcome. For example, an agreed-upon clinical guideline stating that the policy is to treat everyone with a serum cholesterol above the fixed limit—and this might include advice about diet and then recall to the clinic, drugs, further laboratory tests and follow-up, or even simply repetition of the test—results in spending health care resources beyond what might be necessary or expedient. A greater than expected proportion of the population would be labeled "at greater risk," some of whom were misclassified simply due to analytical bias.

In contrast, if the laboratory's bias were negative, the curve would shift to the left. The results are shown in the bottom diagram of Figure 2.3. Some people with serum cholesterol concentration above the fixed limit for clinical action would have low values because of bias. Thus, the number of "false negatives" would increase. This would lead to cost savings on additional testing and drugs in the short term, but potentially huge costs in the long term, as some of the population missed at initial testing succumbed to premature coronary heart disease.

QUALITY SPECIFICATIONS

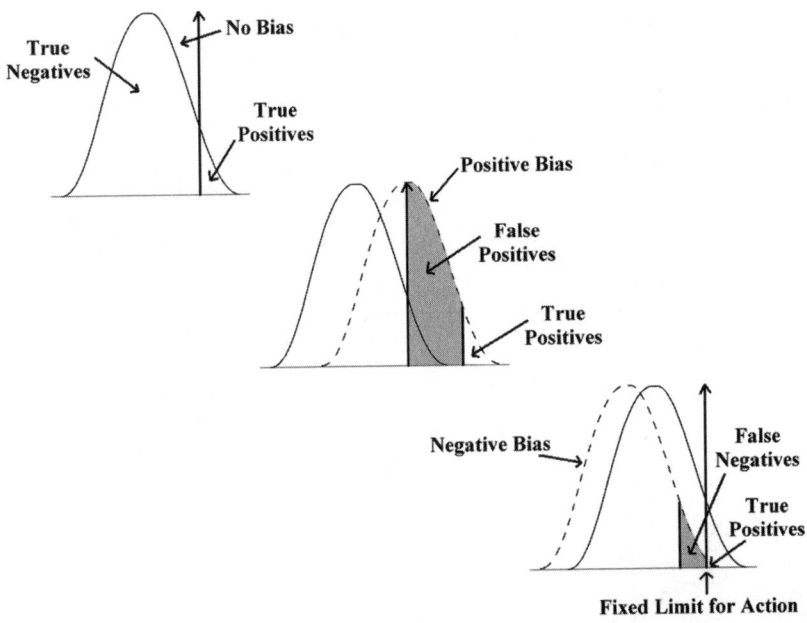

Figure 2.3 The Effect of Bias on Outcomes Using Serum Cholesterol Assays. (Adapted from Hyltoft Petersen P, Horder M. Influence of analytical quality on test results. Scand J Clin Lab Invest 1992;52 (Suppl 208):65–87, with permission.)

The effect of positive and negative biases on the fraction of the population at high risk can be easily derived from knowledge about the simple mathematics of Gaussian distribution: by calculating the percentages of people inside and outside the fixed limit, and doing this for a number of bias values. Then the relationship between analytical bias and the decreases and increases in the percentage of the population at high risk can be calculated and plotted as shown in Figure 2.4.

If we could then define medical needs in terms of percentage misclassification allowable, the analytical bias allowable—the quality specification—can be easily interpolated. In this example, if (and it is a big if) clinicians agreed that it was satisfactory for 5% of people to be incorrectly classified, then we could allow an analytical bias of up to around ± 3–4%.

Note that this approach gives a quality specification for bias. Similar (but more difficult) calculations could be performed to examine the effect of precision on clinical outcomes. However, as we will see later, bias is the most important performance characteristic when using fixed limits for test interpretation.

Identifying such clear clinical strategies as this one is probably the best possible approach to setting quality specifications.

However, a major disadvantage is that most tests results are used in a variety of clinical settings, and only a few tests are used in single well-defined clinical situations with standard well-accepted medical strategies directly related to the test

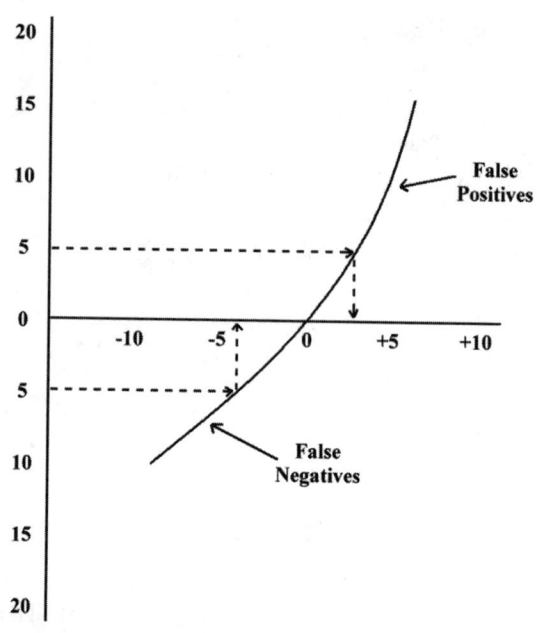

Figure 2.4 Relationship between Positive and Negative Bias and Number of False Positives and Negatives Providing Means of Generating Quality Specifications. *(Adapted from Hyltoft Petersen P, Horder M. Influence of analytical quality on test results. Scand J Clin Lab Invest 1992;52 (Suppl 208):65–87, with permission.)*

result. Another significant drawback is that the calculated quality specifications depend very much on assumptions about how clinicians actually use numerical test results. We have tried asking clinicians how they interpret tests used in very limited clinical situations—such as hemoglobin A_{1c} in monitoring diabetics—but they seem unwilling or unable to actually define, in specific terms, exactly how the test results are used in clinical practice.

Quality Specifications Based on General Clinical Uses of Test Results We know that clinical laboratory test results are used for many purposes. The two major clinical settings for using test results—(1) monitoring individual patients and (2) diagnosis or case finding using reference intervals—show that generally applicable quality specifications are probably best based on components of biological variation, namely, on within-subject and between-subject variation.

The second approach in this group (the second level in the hierarchy) is based on the thesis that we can produce general quality specifications by seeking clinician input. In the past, this has been done in only a few studies, and in general, rather badly. However, the concept does seem good: clinicians use our test results so they should be able to tell us what quality is required. Thus, this strategy could generate quality specifications based upon perceived medical needs. We calculate

QUALITY SPECIFICATIONS

quality specifications based on how clinicians respond to a series of short case studies, or vignettes, on the general interpretation of test results. An example follows.

> A 63-year-old man with high blood pressure has a serum cholesterol concentration of 6.60 mmol/L. He is advised about lifestyle modification including diet. You review him after 2 months.
>
> What serum cholesterol concentration would indicate to you that he has taken your advice?

The optimum approach to surveying clinicians should employ a number of steps.

- Ideally, select a single test and a single major clinical setting for which quality specifications are required.
- Then select a group of clinicians who regularly use the analysis.
- Write a series of case histories that describe common, relatively well-defined clinical situations in which the analyte is a crucial part of patient care.
- Circulate a questionnaire, or present vignettes to clinicians in personal one-to-one interviews.

The case history is about a patient with a well-defined clinical condition. A first result is given for that particular patient. Then, the clinician is asked to give the particular value that is thought to be just different enough from the first value that so as to alter the clinical decision-making. The first value may be a value inside or outside the conventional reference interval, or it may be a population-based reference limit.

Calculating Quality Specifications for Precision from Responses to Clinical Vignettes   It is easy to perform the detailed calculations required for the data analysis (this will be discussed further in Chapter 3). Since we are looking here at changes in a single subject over time, the performance characteristic of importance in this situation is precision rather than bias, although, as we will see later, bias can be included. The steps, as applied to the vignette of the 63-year-old man presented earlier, are as follows.

- Collate responses.
- Calculate differences between 6.60 and responses.
- Calculate frequency distribution of differences.
- Calculate the median, the 25th, and the 75th percentile of differences.
- Decide on the probability of the word *indicate* and find appropriate Z-score (the relationship between semantics, probability, and Z-score will be discussed in Chapter 3).
- Find the within-subject biological variation of cholesterol from the literature as detailed in Chapter 1.

- Calculate the analytical performance required to make this clinical decision at the desired level of probability.
- Use the median, the 25th, and the 75th percentile of differences to create three levels of quality specification: *desirable, optimum,* and *minimum.*

The clinicians have told us what change is thought to be clinically significant. We then take into account the probability, which must be appropriate to the semantics of the question submitted to the clinicians, since different words imply different levels of probability. In addition, given that the differences suggested as significant are based on serial results in an individual, these differences include biological variation. Within-subject biological variation, gleaned from the extensive literature, must be taken into account.

We usually get a wide variety of responses, even for a single analyte in a single clinical setting. We commonly use the median of responses as the desirable quality specification. Optimum and minimum quality specifications can then be defined as the 25th and 75th percentile of the responses, respectively. These quality specifications will generally be concerned with desirable precision.

Sophisticated studies could assume that the change deemed interesting was due to total error and dissect out components due to precision and bias. (We will see how to do this in Chapter 3.)

Most of the vignette-type studies done in the past have been conducted less well than that in the description given above (they have had significant deficiencies in design and execution). Even so, some still quote—quite incorrectly—the quality specifications calculated from those studies. Yet the approach does seem to have considerable potential, and we hope that more studies, properly conducted, will be performed in the future. Recently, a very interesting study from Norway, which approaches the ideal above, actually used diabetic patients who gave opinions on their own self-monitoring of blood glucose. This investigation could be taken as a model for future studies.

Quality Specifications from Professional Recommendations A small number of international and national professional groups have proposed detailed quality specifications. Some of these are concerned with precision, some with bias, and others with total error allowable. Widely used quality specifications based on these proposals include the following.

- The National Cholesterol Education Panel in the U.S. published recommendations for the precision, bias, and total error allowable of lipid analyses.
- The American Diabetes Association documented quality specifications for self-monitoring blood glucose systems and for glycated hemoglobin analyses.
- Expert groups from the National Academy of Clinical Biochemistry in the U.S. proposed quality specifications for thyroid hormone assays, therapeutic drug analyses, and for tests used in the diagnosis and monitoring of diabetes mellitus and liver function. Interestingly, the guidelines on thyroid hormone assays are under review and the new guidelines suggest that quality specifica-

tions for precision, bias, and total error allowable are best based on components of biological variation, as are certain of the guidelines on diabetes and liver function.
- A European Working Group has proposed quality specifications for use in the evaluation of the precision and bias of analytical systems, again based on components of biological variation.
- Another European Working Group has suggested quality specifications for reference methods for validating routine methods and for assigning values to materials used in PT or EQAS, again based on components of biological variation.

These quality specifications are based on the considerable laboratory and clinical experience of the people involved in their generation, and usually on detailed discussion of the available evidence before publication. Users of the specifications can evaluate the objectivity of the process used to reach the conclusions because the method by which the recommendations were reached is published in the literature.

Steps in Preparing a Consensus Document A proposed strategy (not all steps need be included) for deriving guidelines for quality specifications using the expert professional recommendations approach is as follows.

- Professional body decides need and appoints expert panel.
- Expert panel decides scope of recommendation.
- Professional body agrees on scope and approves further work.
- Experts write document components.
- External peers review document components.
- Document is collated.
- Document is presented at conference (and on Internet) for response.
- Document is modified.
- Redrafted document is reviewed by external peers.
- Redrafted document is posted on Internet for review.
- Appropriate views are taken into account.
- Final document is prepared.
- Final document is published in whole in appropriate journals.
- Executive summary is published widely.
- Document is reviewed at stated future time.

Less widely used are those quality specifications that have been proposed in published guidelines—"best practice" or "good laboratory practice" guidelines. These are often developed or presented at a single consensus conference without significant discussion. They have some merit in that they are usually based on the wide knowledge of an expert or an expert group from a single institution. However, the guidelines are often subjective and are not often based on accepted models, new approaches, or experimental data. These quality specifications are further down the hierarchy than proposals from national or international expert groups.

Because quality specifications are of very different types—some give data for precision, bias, and total error allowable separately; others give data on only one or two of these characteristics—careful reading of the recommendations before they are applied inappropriately is highly recommended.

Quality Specifications Based on Regulation and External Quality Assessment

A few countries have defined analytical performance standards that laboratories must meet in order to be considered acceptable, or in some cases, to achieve and/or retain accredited status. The U.S. CLIA'88 legislation documents total error allowable, which is precision plus bias, of course, for a number of commonly assayed analytes. Some are shown in Table 2.2. Similar legislation exists in Germany but the quality specifications are very different from those in the U.S. (for example, German Federal law requires that precision (CV_A) be < 1/12 of the reference interval).

The advantage of this strategy is that the CLIA'88 quality specifications are well known and understood and widely available even on the Internet (www.westgard.com/clia.htm). However, a major disadvantage is that the CLIA'88 quality requirements appear to be based upon what is achievable rather than what is desirable. Moreover, when legislation exists and lays down "acceptable" standards of performance, then laboratories might see achievement of these as the desirable goals rather than use any other set of quality specifications. Much recent literature on quality planning uses the CLIA '88 set of quality specifications for total error allowable as the basis for the model.

The many different EQAS in use all over the world use different techniques to judge the acceptability or other performance criteria attained by the participant laboratories. Some countries analyze data returned from participant laboratories, applying the overall or method group consensus mean to assess bias and using the found SD or CV to create a window of acceptability, commonly as 3SD or 3CV. This has significant disadvantages because the SD or CV merely shows what is achievable with current methodology and technology.

Thankfully, it appears that more laboratory professionals are using fixed limits as criteria of acceptability. These, like the CLIA '88 criteria, generally denote total error allowable. The major disadvantage of using these EQAS fixed limits as quality specifications is that, although often based on expert opinion, they tend to be empirical. Different countries use quite different fixed limits, which tends to support the view that they are not totally objective. They also are clearly much influenced by what is actually achievable with current technology and methodology or, as it is termed, "the state of the art."

In spite of these difficulties, the state of the art as evidenced from PT or EQAS has been advocated quite widely in the past as quality specifications, particularly when the performance achieved by better laboratories, typically the best 20%, is taken as the target. The underlying concept is that, if one in five laboratories can achieve this level of quality, then the technology and methodology exist for all laboratories to achieve the same level of analytical performance.

QUALITY SPECIFICATIONS

Table 2.2 Examples of CLIA '88 Quality Specifications for Acceptable Performance

Analyte	Acceptable Performance
ALT	Target value ± 20%
Albumin	Target value ± 10%
Alk Phos	Target value ± 30%
Amylase	Target value ± 30%
AST	Target value ± 20%
Bilirubin	Target value ± 0.4 mg/dL or ± 20%
Calcium	Target value ± 1.0 mg/dL
Chloride	Target value ± 5%
Cholesterol	Target value ± 10%
HDL-Cholesterol	Target value ± 30%
CK	Target value ± 30%
Creatinine	Target value ± 0.3 mg/dL or ± 15%
Glucose	Target value ± 6 mg/dL or ± 10%
Iron	Target value ± 20%
LD	Target value ± 20%
Magnesium	Target value ± 25%
Potassium	Target value ± 0.5 mmol/L
Sodium	Target value ± 4 mmol/L
Total protein	Target value ± 10%
Triglycerides	Target value ± 25%
BUN (urea nitrogen)	Target value ± 2 mg/dL or ± 9%
Urate	Target value ± 17%
Hematocrit	Target value ± 6%
Hemoglobin	Target value ± 7%
Leukocyte count	Target value ± 6%
Erythrocyte count	Target value ± 15%
Platelet count	Target value ± 25%
Fibrinogen	Target value ± 20%
PTT	Target value ± 15%
PT	Target value ± 15%

Quality Specifications Based on the State of the Art Data about what is actually achieved analytically are often available from PT and EQAS organizers; if nothing else were available as a quality specification, we could use this generally attainable state of the art. However, the documented analytical performance may not truly reflect the state of the art because samples circulated to participating laboratories may not behave exactly like patient samples, due to matrix effects. In addition, laboratory staff may handle these samples with special care to try to "improve" their performance. The state of the art documented in PT and EQAS changes with time (and not always for the better), and the performance achieved may bear no relationship to actual medical needs.

One can glean the state of the art by reading original publications on methodology in peer-reviewed literature. There is a caveat: performance documented in the laboratory of the originator or the original evaluator may be the best possible (because of operating under close to ideal conditions) rather than what is achieved in everyday practice. Again, performance achieved analytically may bear no relationship to actual medical needs.

Thus, these approaches appear low in the hierarchy and certainly below quality specifications based on biological variation.

STRATEGIES FOR SETTING QUALITY SPECIFICATIONS BASED ON BIOLOGICAL VARIATION

All strategies for setting quality specifications such as precision, bias, and total error allowable in laboratory medicine have advantages and disadvantages. The fundamental principle, of course, is that quality specifications should be

- firmly based upon medical requirements,
- usable in all laboratories irrespective of size, type or location,
- generated using simple to understand models, and
- widely accepted as cogent by professionals in the field.

Quality specifications based on biology seem to fulfill all these criteria and will be examined in detail in this section.

Uses of Clinical Laboratory Test Results Laboratory test results are used for many purposes. We use them in teaching and training, and in research and development projects ranging from the very basic to the very applied. We also use test results clinically mainly in four rather different situations.

Diagnosis involves identifying disease by investigating symptoms, and this usually involves performing a range of relevant clinical laboratory tests.

Case finding is the opportunistic performance of a panel of investigations, usually including a range of clinical laboratory tests, when an individual presents to the health care system.

Screening is the identification of unrecognized disease or defect, and is applied to the apparently healthy.

Monitoring involves reviewing laboratory test results over time. The time period may be short (for example, during an acute disease episode treated in the hospital); medium-term (for example, measuring tumor markers to assess recurrence); or long term (for example, monitoring glycemic control in diabetes mellitus).

Quality specifications for precision and bias should ensure that these clinical purposes can be achieved. If we can develop separate quality specifications for precision and bias, it is easy to calculate specifications for total error allowable.

Quality Specifications for Precision: Calculating Total Variation Random variation, or precision, is defined as the closeness of agreement between inde-

QUALITY SPECIFICATIONS

pendent results of measurements obtained under stipulated conditions. In practice, precision is measured by replicating analysis of the same sample in our internal quality control program.

To answer the question, "How low should precision be?", we must ask, "What effects does precision have on test results and clinical decision-making?".

Before we can investigate this numerically, we must explore the calculation of total variation more objectively and mathematically. There are two general formulae that are relevant in this context.

First, if the test result is calculated by addition or subtraction, then the total variance is the sum of the variances in standard deviation terms, that is,

If $C = A + B$ or if $C = A - B$, and the measurements of A and B have analytical precisions of SD_A and SD_B respectively,
then, $SD_C^2 = SD_A^2 + SD_B^2$ so that $SD_C = (SD_A^2 + SD_B^2)^{1/2}$.

An example would be the calculation of the "anion gap" which is

anion gap = (sodium + potassium) − (chloride + bicarbonate).

Now, if the SD of sodium analyses were 1.0 mmol/L, potassium 0.1 mmol/L, chloride 1.0 mmol/L, and bicarbonate 0.5 mmol/L, then the SD of the anion gap estimation would be equal to

$$(1.0^2 + 0.1^2 + 1.0^2 + 0.5^2)^{1/2} = (1.00 + 0.01 + 1.00 + 0.25)^{1/2} = 2.26^{1/2} = 1.50$$

Note that the resultant SD numerically exceeds any of the component SD but is not the same as simple arithmetic addition of the component SD; addition must be done as variances.

When all the components have the same mean—and this is a very important proviso—then, and only then, can CV be substituted in the formula for SD.

Second, if quantity is calculated by multiplication or division, then the total variance is the sum of the variances. But this must be done in CV terms, i.e.,

If $C = A * B$ or if $C = A/B$, and the measurement of A and B have analytical precisions of CV_A and CV_B respectively,
then, $CV_C^2 = CV_A^2 + CV_B^2$ so that $CV_C = (CV_A^2 + CV_B^2)^{1/2}$.

As we have discussed in detail in Chapter 1, all analytes assayed in the clinical laboratory vary inherently due to

- pre-analytical variation,
- analytical variation, and
- within-subject biological variation.

These variations are all random. Thus, they can be considered to have Gaussian distributions. As we have seen, the dispersion (width, magnitude) of a Gaussian distribution can be described in terms of standard deviation (SD). At this stage, to keep the analysis as simple as possible, we will consider that pre-analytical variation is negligible—and will explore how to achieve this in detail in Chapter 3.

If analytical variation is called SD_A and within-subject biological variation is called SD_I then total variation (SD_T) can be calculated as follows:

$$SD_T^2 = SD_A^2 + SD_I^2 \text{ or } SD_T = (SD_A^2 + SD_I^2)^{1/2}.$$

If we determine or estimate CV_A at the same level as CV_I, the means of the values will be the same in this case, so the calculation of total variation becomes

$$CV_T^2 = CV_A^2 + CV_I^2$$

or

$$CV_T = (CV_A^2 + CV_I^2)^{1/2}.$$

The Effect of Precision on Test Result Variability We report the results of our analyses as single numbers, but each number has an inherent variation. If we ignore pre-analytical variation, this variation is then due to within-subject biological variation and analytical random variation—precision and changes in bias (due to calibration changes, for example) which we usually include in our precision estimates and which we should minimize as much as possible. Thus, since we now know that we can consider that within-subject biological variation is a constant, the amount of analytical "noise" added to our biological "signal" depends only on the analytical precision.

We can calculate the effect of changing precision on our inherent variation. We know that

$$CV_T = (CV_A^2 + CV_I^2)^{1/2}.$$

Thus, if the analytical precision were exactly the same magnitude as within-subject biological variation, that is, the signal and the noise were exactly equal, then $CV_A = CV_I$ and so, by simple substitution in the formula,

$$CV_T = (CV_I^2 + CV_I^2)^{1/2} = (2CV_I^2)^{1/2} = 1.414\ CV_I$$

which means that intrinsic variation (due to biology) has increased 41.4% because of analytical variation. True test result variability has been increased by 41.4% due to analysis.

QUALITY SPECIFICATIONS

Similarly, if precision were twice as large as within-subject biological variation,

$$CV_A = 2\,CV_I$$

and so

$$CV_T = [(2CV_I)^2 + CV_I^2]^{1/2} = (4CV_I^2 + CV_I^2)^{1/2} = (5CV_I^2)^{1/2} = 2.236\,CV_I$$

which means that the intrinsic variation (due to biology) has increased 123.6% because of analytical variation. True test result variability has been increased by 123.6% due to analysis.

On the other hand, if precision was only half as large as within-subject biological variation,

$$CV_A = \tfrac{1}{2} CV_I$$

and so

$$CV_T = [(\tfrac{1}{2}CV_I)^2 + CV_I^2]^{1/2} = (\tfrac{1}{4}CV_I^2)^{1/2} = (\tfrac{5}{4}CV_I^2)^{1/2} = 1.118\,CV_I$$

which means that intrinsic variation (due to biology) has increased 11.8% because of analytical variation. True test result variability has been increased by 11.8% due to analysis.

We can do similar calculations about how much true test result variability has been increased by analysis for a large range of precision values. Table 2.3 shows these values.

The relationship between amount of variation added to true test result variability to the ratio CV_A/CV_I is not linear. As precision increases, the amount of analytical "noise" added to the biological "signal" increases relatively more. It should be noted that this holds true especially once the precision is numerically larger than the within-subject biological variation.

The Effect of Precision on Variability of a Cholesterol Result Increasing precision—i.e., deteriorating test performance—increases the amount of test result variability. Let us now put the theory discussed above into a clinical context.

Our 63-year-old man with hypertension (the subject of the earlier vignette) has a serum cholesterol concentration of 6.60 mmol/L. We know that the within-subject biological variation of cholesterol is 6.0%. Thus, the inherent variation of the man's serum cholesterol is 6.0% in CV terms or 0.40 mmol/L in SD terms.

Thus, since we know, from the Gaussian distribution characteristics, that:

Table 2.3 Amount of Variation Added to True Test Variability as Precision Becomes Larger Compared to Within-Subject Biological Variation

Ratio of precision to within-subject biological variation (CV$_A$/CV$_I$)	Amount of variation added to true test result variability (as percentage of true variation)
0.25	3.1
0.50	11.8
0.75	25.0
1.00	41.4
1.50	80.3
1.73	100.0
2.00	123.6
2.50	169.3
3.00	216.2
4.00	312.3
5.00	409.9

- mean ± 1SD encompasses 68.3% of the results,
- mean ± 2SD encompasses 95.5% of the results, and
- mean ± 3SD encompasses 99.7% of the results,

then, from a purely biological perspective,

- there is a 68.3% probability that the value lies within 6.60 ± 0.40 mmol/L = 6.20–7.00 mmol/L,
- there is a 95.5% probability that the value lies within 6.60 ± 0.80 mmol/L = 5.80–7.40 mmol/L, and
- there is a 99.7% probability that the value lies within 6.60 ± 1.20 mmol/L = 5.40–7.80 mmol/L.

If the analytical precision were 3%, as recommended by the National Cholesterol Education Program in the U.S., then the total variation would be

$$CV_T = (CV_A^2 + CV_I^2)^{1/2}$$
$$= (6^2 + 3^2)^{1/2}$$
$$= 6.7\%$$

so that there is a 95.5% probability that the cholesterol lies within 6.60 ± 0.88 mmol/L = 5.72–7.48 mmol/L.

If precision is 5%, there is a 95.5% probability that the cholesterol lies within 6.60 ± 1.03 mmol/L = 5.57–7.63 mmol/L.

If precision is 10%, there is a 95.5% probability that the cholesterol lies within 6.60 ± 1.54 mmol/L = 5.06–8.14 mmol/L.

QUALITY SPECIFICATIONS

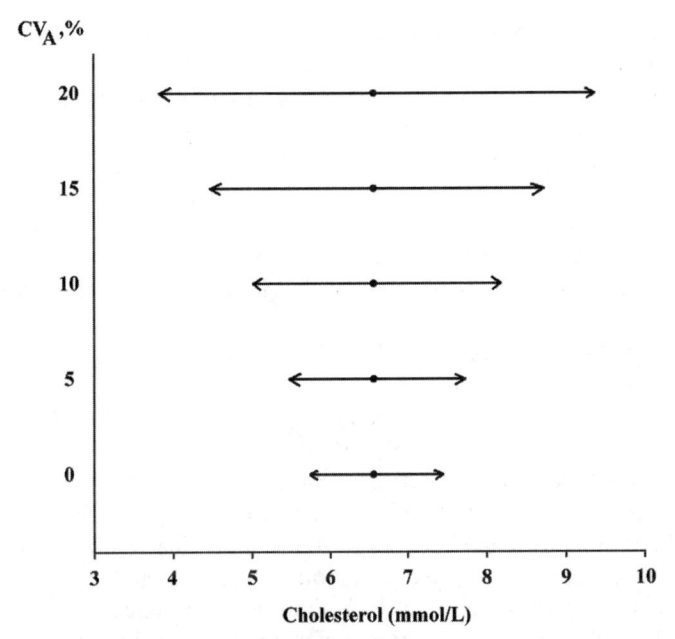

Figure 2.5 Dispersion (95.5%) for Serum Cholesterol of 6.60 mmol/L at Various Levels of Analytical Precision

The range of the 95.5% dispersion of the single cholesterol result with increasing precision is shown in Figure 2.5. Note again the non-linear nature of the influence of deteriorating precision. The diagram is not an isosceles triangle with straight sides—the sides are concave towards the center. Ever-worse precision gives ever-bigger dispersion.

We have already seen that change in serial results in an individual over time is due to pre-analytical variation, analytical variation (precision and changes in bias), and within-subject biological variation. Thus, again, since errors are additive, inferior precision will make it harder to monitor people over time since large changes may be simply due to analytical variation rather than to a truly signifying improvement or deterioration. Clinical "signal" is drowned out by analytical "noise." This vitally important influence of precision on interpretation of serial results in monitoring individuals will be covered in detail in Chapter 3.

Population-based reference values (which will be discussed in detail in Chapter 4) are very frequently used as aids to interpretation. The reference interval is calculated from results obtained on samples from reference individuals. Every one of these results contains a component of variation due to analytical precision. Clearly, values developed using a method with poor precision will have a broader reference interval than values generated for the same analyte using a method with good precision. Broader reference intervals confounded by analytical variation will be of less clinical utility because individuals will be classified incorrectly more often.

Quality Specifications for Precision Based on Biological Variation Low precision reduces inherent variability for every individual test result. (We will explore later how low precision leads to greater probability of significance for changes in serial results from an individual, and to narrower population-based reference intervals, leading to better diagnostic accuracy.)

If we know precision is low, we will be able to run fewer internal quality control samples per analytical run and/or we will be able to use less stringent quality control rules. We will increase the probability for error detection and decrease the probability of falsely rejecting the results. This is an important concept for quality planning.

But the vital question remains: *How low is good enough?* We know that increasing precision leads to increasing amount of variability added to test result variability. We calculated in detail that, as CV_A rises, the amount of added variation rises—and this rise is not simply linear.

The concept that analytical variation should be less than one-half the average within-subject biological variation is not new and was developed some 30 years ago. We have already calculated that, if the analytical variation is less than one-half the average within-subject variation, then the amount of variability added to true test result variability is about 10%. Only 10% analytical "noise" is added to the true biological "signal." This amount of added analytical variability seems reasonable (although it must be admitted that this is a rather empirical judgement) and leads us to postulate that the best quality specification for precision is

analytical precision < one-half the within-subject biological variation, or
$CV_A < 0.50 CV_I$.

This model rates very highly in the hierarchy of quality specifications, second only to assessment of the effect of analysis on clinical decision making. In view of the many difficulties of the analysis of outcome approach, quality specifications based on components of biological variation are much favored and widely used, indeed, for many years. Using them is easy because estimates of within-subject biological variation are constant over time and geography. In addition, readily available data on average within-subject biological variation makes it easy to calculate quality specifications. Moreover, many quality specifications proposed in international and national guidelines—level 3 in the hierarchy—are also based on biological variation.

This basic concept has been expanded: increasing the analytical precision relative to the within-subject biological variation increases the test result variability. We showed earlier the simple calculations that would allow us to determine that:

- When $CV_A < 0.75 CV_I$, then at most 25% variability is added to test result variability.
- When $CV_A < 0.50 CV_I$, then no more than 12%% variability is added.
- When $CV_A < 0.25 CV_I$, then a maximum of 3% variability is added.

QUALITY SPECIFICATIONS

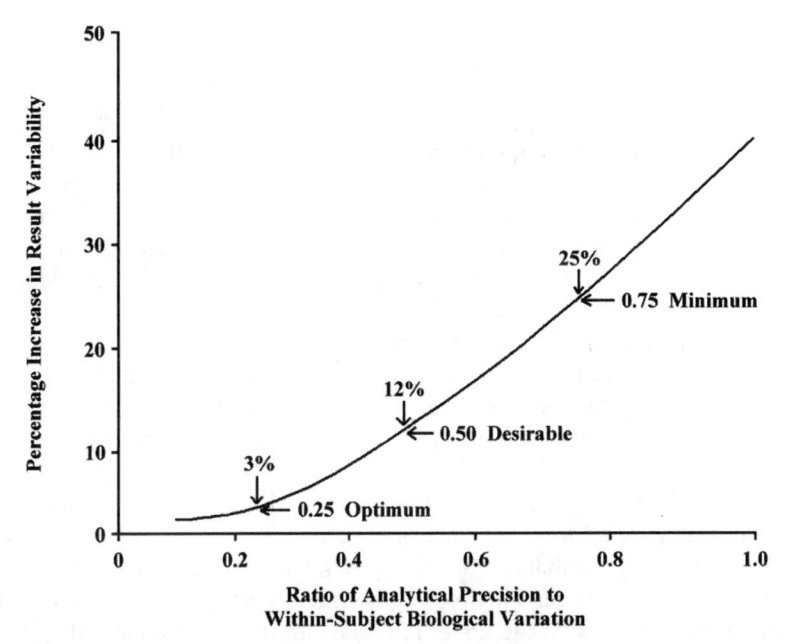

Figure 2.6 Quality Specifications for Precision Showing the Amount of Test Result Variability Added As a Function of the Ratio of Precision to Within-Subject Biological Variation. (*Adapted, with permission, from Fraser CG, et al. Proposals for setting general applicable quality goals solely based on biology. Ann Clin Biochem 1997;34:8–12.*)

It has been proposed, therefore, as shown in Figure 2.6, that:

- *Desirable performance* is defined by $CV_A < 0.50 CV_I$. Quality specifications generated using this formula should be viewed as being generally applicable. This is the original, most widely accepted, and very frequently used quality specification based on biological variation, but we have suggested that, in order to cater to those analytes for which the general quality specifications appear too "loose" or too "stringent," then:
- *Optimum performance* is defined by $CV_A < 0.25 CV_I$. The more-stringent quality specifications generated using this formula should be used for quantities for which desirable performance standards are easily achieved with current technology and methodology.
- *Minimum performance* is defined by $CV_A < 0.75 CV_I$. The less-stringent quality specifications generated using this formula should be used for those quantities for which the desirable performance standards are not attainable with current technology and methodology.

Desirable quality specifications for precision for a wide variety of analytes are shown in Appendix 2. The biological variation of these, shown in Appendix 1, can be used when required to calculate *optimum* and *minimum* quality specifications for precision as and when appropriate.

Influence of Performance on Reference Values   Clearly, the dispersion of the reference interval will depend on the precision of the analytical procedure. As we have seen, the worse the precision, the wider the reference interval. We can calculate this quite easily using addition of variances as demonstrated earlier.

However, bias is more important. The reference limits will be very dependent on the analytical bias. This is illustrated in Figure 2.7.

The top diagram shows an error-free Gaussian distribution. By definition—and by current convention—reference limits are set to ensure that 95% of population values lie within the reference interval. Thus, 2.5% of the group will have values above the upper reference limit and 2.5% will have values below the lower reference limit.

Now, if a method has positive bias, the curve will move to the right as shown in the middle diagram. More that 2.5% of the group will have values above the upper reference limit. Less than 2.5% of the group will have values below the lower reference limit. It is important to note that, because of the bell-shaped distribution, the increase above 2.5% at the upper reference limit will be greater than the decrease below 2.5% at the lower reference limit.

Another way of thinking about the effect of this positive bias is that there will be more clinical false positives than false negatives. The important end result is that more than 5% of people will be classified as unusual—more than the desired 5% will have values outside the reference interval.

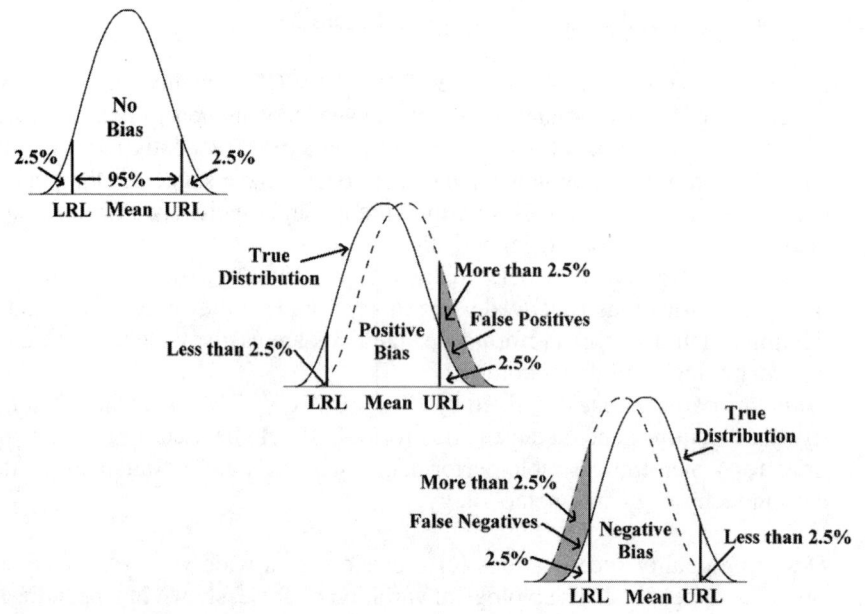

Figure 2.7   Effect of Bias on Reference Values

QUALITY SPECIFICATIONS   **53**

Similarly, if a method has negative bias, the curve will move to the left as shown in the bottom diagram. More that 2.5% of the group will have values below the lower reference limit. Less than 2.5% of the group will have values above the upper reference limit. It is again important to note that, because of the bell-shaped distribution, the 2.5% decrease below the lower reference limit will be greater than the 2.5% increase above the upper reference limit.

The other way of thinking about the effect of negative bias is that there will again be more erroneous results outside the lower reference limit than errors inside the upper limit. The result is again that more than 5% of people will be classified as unusual—more than the desired 5% will have values outside the reference interval.

Quality Specifications for Bias Based on Biological Variation   A positive bias will increase the percentage outside the upper reference limit and decrease the percentage outside the lower reference limit. A negative bias will have the same effects but on the opposite reference limits. From the mathematics of the Gaussian distribution, we can calculate how many people will be outside each reference limit when bias exists.

From a medical point of view, then, the fundamental concept is for laboratories throughout a homogeneous population area to use the same reference intervals. This means that laboratory data would be transferable between laboratories, so that patients would not have to get their laboratory tests repeated every time they visit a different hospital. Even if patients attended different general practitioners who used different laboratories, laboratory results would be comparable if they had only a small bias. In addition, when laboratories change analytical systems or methods, the ideal would be that the reference values used by the laboratory—which take much time, effort and resources to generate—would be able to be continued without modification.

But how much bias can we allow so that reference intervals are transferable over geography and time?

The reference interval is made up of within-subject biological variation (CV_I) and between-subject biological variation (CV_G), and, if analytical precision is considered negligible, this "group" biological variation can be calculated, again as simple addition of variances, as $(CV_I^2 + CV_G^2)^{1/2}$. Remember that we can use CV in this formula because the mean values of the components are the same.

For us all to use the same set of reference values, then, analytical bias should be less than one-quarter the group biological variation, or

$$B_A < 0.250(CV_I^2 + CV_G^2)^{1/2}.$$

We can calculate that when $B_A < 0.250\,(CV_I^2 + CV_G^2)^{1/2}$, then 1.4% are outside one reference limit and 4.4% outside the other. Thus, less than 1% more (0.8%) of the group are outside the reference interval than the 5% we would expect by definition. The increase in numbers of people outside the reference interval is 0.8/5.0 = 16% and, analogously with the setting of quality specifications for desirable precision, this seems "reasonable" for a general quality specification.

We can also calculate that, when $B_A < 0.375(CV_I^2 + CV_G^2)^{1/2}$ then 1.0% are outside one reference limit and 5.7% outside the other, so that about 1.7% more than the desirable 5% are outside the reference interval (an increase in numbers of people outside the reference interval of 1.7/5.0 = 34%).

When $B_A < 0.125 (CV_I^2 + CV_G^2)^{1/2}$, then 1.8% are outside one reference limit and 3.3% outside the other, so that about 0.1% more than the desirable 5% are outside the reference interval (an increase in numbers of people outside 0.1/5.0 = 2%).

This infers that, exactly like precision, we should have three levels of quality specification, as shown in Figure 2.8.

- *Desirable performance* is defined by $B_A < 0.250 (CV_I^2 + CV_G^2)^{1/2}$. The quality specifications generated using this formula should be viewed as those generally applicable. This is the original, most widely accepted, and very frequently used quality specification based upon biological variation, but we

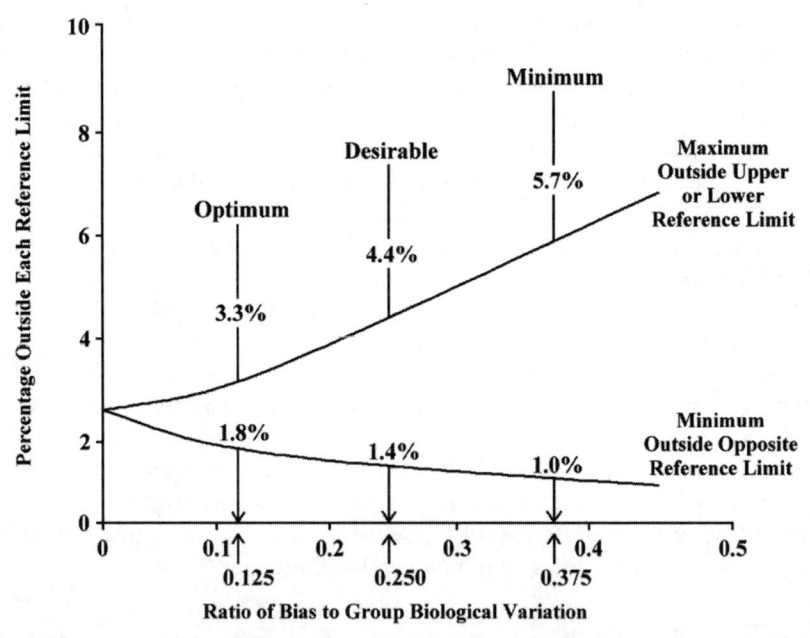

Figure 2.8 Quality Specifications for Bias Showing the Percentage of the Population Outside the Reference Limits as a Function of the Ratio of Bias to Group Biological Variation. (*Adapted with permission from Fraser CG, et al. Proposals for setting general applicable quality goals solely based on biology. Ann Clin Biochem 1997;34:8–12.*)

QUALITY SPECIFICATIONS

have suggested that, in order to cater to those analytes for which the general quality specifications appear too "loose" or too "stringent," then:

- *Optimum performance* is defined by $B_A < 0.125 \, (CV_I^2 + CV_G^2)^{1/2}$. The more-stringent quality specifications generated using this formula should be used for quantities for which desirable performance standards are easily achieved with current technology and methodology.
- *Minimum performance* is defined by $B_A < 0.375 \, (CV_I^2 + CV_G^2)^{1/2}$. The less-stringent quality specifications generated using this formula should be used for those quantities for which the desirable performance standards are not attainable with current technology and methodology.

The *desirable* quality specifications for a wide variety of analytes are shown at Appendix 2. The biological variation of these, shown in Appendix 1, can be used when required to calculate *optimum* and *minimum* quality specifications for bias as and when appropriate.

Quality Specifications for Total Error Allowable It is widely accepted that quality specifications are best based on biological variation, the second level of quality specification in the hierarchy of models, so since the general desirable quality specifications are

$$CV_A < 0.50 CV_I$$
$$B_A < 0.250 \, (CV_I^2 + CV_G^2)^{1/2}$$

then the *desirable* quality specification for total error allowable (using the formula derived earlier) is

$$TE_a < 1.65(0.50 \, CV_I) + 0.250 \, (CV_I^2 + CV_G^2)^{1/2}$$

The "three-level model" allows for analytes that cannot meet these general quality specifications using current methodology and technology, for example, assays of calcium and sodium in serum. For these difficult analyses

$$CV_A < 0.75 CV_I$$
$$B_A < 0.375 \, (CV_I^2 + CV_G^2)^{1/2}$$

and thus the *minimum* quality specification for total error allowable is

$$TE_a < 1.65(0.75 \, CV_I) + 0.375 \, (CV_I^2 + CV_G^2)^{1/2}$$

For example, for chloride, $CV_I = 1.2\%$ and $+ CV_G = 1.5\%$, so that the *desirable* quality specifications are

$CV_A < 0.50 CV_I = 0.6\%$
$B_A < 0.250 \, (CV_I^2 + CV_G^2)^{1/2} = 0.250 \, (1.2^2 + 1.5^2)^{1/2} = 0.5\%$

and

$TE_a < 1.65(0.50 \, CV_I) + 0.250 \, (CV_I^2 + CV_G^2)^{1/2} = 1.65 \, (0.6) + 0.5 = 1.5\%$

It might be that these somewhat demanding quality specifications could not be met in the laboratory and, while desirable quality specifications should be taken as targets worthy of achievement when methodology and technology allow, it might be better to have realistic specifications for use in quality planning and management. These would be based upon the minimum quality specifications formulae:

$CV_A < 0.75 CV_I = 0.9\%$
$B_A < 0.375 \, (CV_I^2 + CV_G^2)^{1/2} = 0.375 \, (1.2^2 + 1.5^2)^{1/2} = 0.7\%$

and

$TE_a < 1.65(0.75 \, CV_I) + 0.375 \, (CV_I^2 + CV_G^2)^{1/2} = 1.65(0.9) + 0.7 = 2.2\%$

It also allows for analytes that can easily meet general quality specifications with current methodology and technology, for example, assays of triglycerides and CK activity in serum. For these easy analyses,

$CV_A < 0.25 CV_I$
$B_A < 0.125 \, (CV_I^2 + CV_G^2)^{1/2}$

and thus the *optimum* quality specification for total error allowable is

$TE_a < 1.65 \, (0.25 \, CV_I) + 0.125 \, (CV_I^2 + CV_G^2)^{1/2}$

For example, for urea, $CV_I = 12.3\%$ and $+ CV_G = 18.3\%$, so that the *desirable* quality specifications are

$CV_A < 0.50 CV_I = 6.2\%$
$B_A < 0.250 \, (CV_I^2 + CV_G^2)^{1/2} = 0.250 \, (12.3^2 + 18.3^2)^{1/2} = 5.5\%$

and

$TE_a < 1.65(0.50 \, CV_I) + 0.250 \, (CV_I^2 + CV_G^2)^{1/2} = 1.65 \, (6.2) + 5.5 = 15.7\%$

QUALITY SPECIFICATIONS

It is highly possible that these not-very-demanding quality specifications could be met in all laboratory settings and, it might be better to have more stringent specifications for use in quality planning and management. These would be based upon the optimum quality specifications formulae:

$CV_A < 0.25 CV_I = 3.1\%$
$B_A < 0.125 (CV_I^2 + CV_G^2)^{1/2} = 0.125 (12.3^2 + 18.3^2)^{1/2} = 2.8\%$

and

$TE_a < 1.65(0.25\ CV_I) + 0.125 (CV_I^2 + CV_G^2)^{1/2} = 1.65(3.1) + 2.8 = 7.9\%$

Desirable quality specifications for a wide variety of analytes are shown in Appendix 2. The biological variation of these, shown in Appendix 1, can be used when required to calculate *optimum* and *minimum* quality specifications for total error allowable as and when appropriate.

Other Quality Specifications Based on Biological Variation A number of other quality specifications, less widely used but still interesting, can be derived using the available data on components of biological variation.

Sometimes, the same laboratory uses different techniques to analyze a single analyte. Examples are STAT and routine sections, routine and back-up systems, and laboratory and point-of-care test (POCT) analyzers. Often the same patient has samples assayed by these different systems. The systems may have different precision and different bias.

Modern thinking is that, if we have a known bias, it should be eliminated before the results are reported (this is good scientific practice and advocated by international bodies such as IFCC and IUPAC). Moreover, one system (usually the routine system) is made the "gold standard," and the calibration of all other systems is tied into it, if possible.

It is likely, however, that the methods will have their inherent precision, some bias, and changes in bias. In consequence, it is important to ensure that an individual's results are comparable. Often we analyze the same internal quality control samples on different systems (but we use different numbers per run and different rules for acceptance/rejection). A mathematical model based on investigating significant changes in serial results has shown that quality specifications can be set for the allowable difference between two methods used to analyze the same analyte in a single laboratory, as

allowable difference $< 0.33 CV_I$.

For many analytes, this level of performance is attainable with current methods and technology. For example, CV_I for urea is 12.3%, so the allowable difference between methods is 4.1%. In Dundee, the core laboratory has three Hitachi 917 analytical systems, and the differences between the monthly quality control means are always <1.0% for this analyte. In contrast, even though the difference between the means for our sodium analyses differ by only 0.4 mmol/L, our performance does not quite meet the quality specification of allowable difference of 0.2%, since CV_I for sodium is only 0.7%.

When the analytical systems have quite different methodology and calibration techniques, it will be more difficult to meet these sometimes-demanding quality specifications in practice. This is one reason why many laboratories have eliminated separate STAT facilities and use good workflow management to fast-track, rapid response samples through their core laboratory systems, thereby facilitating transfer of laboratory results across geography. Moreover, doing as little point-of-care testing (POCT) as possible through adopting techniques such as vacuum tube delivery systems (to cut turn-around times) also eliminates many problems related to comparison of results obtained on intrinsically different analytical systems.

Quality specifications for drug levels in therapeutic drug monitoring (TDM) can be calculated using a model similar to the one based on biological variation. The model assumes that the steady-state fluctuation in drug concentration between the minimum and maximum is the "biological variation" in this setting. It also assumes that there is negligible bias. Using simple pharmacokinetic theory, the quality specification for precision for TDM is

$$CV_A < 0.25[(2^{T/t}-1)/(2^{T/t}+1)] * 100,$$

where T is the dosing interval and t the half-life. This model seems cogent with the change in drug concentrations with time. Stringent quality specifications are derived for drugs with short dosing intervals and long half-lives—and these are the drugs that vary little in steady state.

Digoxin is usually given in a single daily dose. In individuals whose renal function is not impaired, the average half-life is 38.4 hours. From the model, the desirable precision is thus

$$CV_A < 0.25[(2^{24/38.4}-1)/(2^{24/38.4}+1)] * 100 = 0.25[(1.54-1)/(1.54+1)] * 100$$
$$= 5.3\%.$$

A drug with a shorter half-life is carbamezepine (average half-life of 16 hours) but this drug is usually given twice each day. It would be expected, from the model, that the quality specification would be similar to digoxin because, although the half-life is less, the dosing interval is less, too.

$$CV_A < 0.25[(2^{12/16}-1)/(2^{12/16}+1)] * 100 = 0.25[(1.68-1)/(1.68+1)] * 100$$
$$= 6.4\%.$$

A Working Group of the European EQAS Organizers Group has investigated objective means to set quality specifications for application as the fixed limits of

QUALITY SPECIFICATIONS

Table 2.4 Proposed European Quality Specifications for TE_a in PT and EQAS

Analyte	TE_a for PT and EQAS
Sodium	0.9
Potassium	7.2
Calcium	2.8
Magnesium	4.2
Glucose	7.0
Creatinine	7.9
Cholesterol	10.4
Urea	20.8

acceptance in PT and EQAS. The model uses exactly the quality specification for TE_a (for 99% probability), i.e.:

allowable error $< 0.25(CV_I^2 + CV_G^2)^{1/2} + 2.33\,(0.05 CV_I)$.

This is a simple combination of the desirable quality specifications for precision and bias—exactly as for calculating TE_a shown earlier. Some of the quality specifications advocated by the European Group are shown in Table 2.4.

Another European Working Group has addressed the question of quality specifications for reference methods. It was suggested that, when such methods were used to validate routine methods, the quality specifications based on biology proposed earlier should be used but that, for this application, they be halved. They then become exactly the *minimum* set of quality specifications based on biology:

$CV_A < 0.25 CV_I$

and

$B_A < 0.125(CV_I^2 + CV_G^2)^{1/2}$.

However, when the methods were used to set values for use in EQAS, the fixed limits of acceptance should be the 99% limits for TE_a but these should be divided by a factor of five:

$TE_a < 0.20[0.25(CV_I^2 + CV_G^2)^{1/2} + 2.33\,(0.05 CV_I)]$

Quality specifications for the TE_a of reference methods in these two rather different applications are shown in Table 2.5.

It appears clear, then, that the professional consensus is that quality specifications are best based on calculations involving components of biological variation.

Table 2.5 Quality Specifications for Reference Methods

Analyte	TE$_a$ for use in validation of routine methods studies	TE$_a$ for use when assigning values to materials circulated in EQAS
Sodium	0.1	0.2
Potassium	0.8	1.4
Calcium	0.4	0.6
Magnesium	0.6	0.8
Glucose	1.0	1.4
Creatinine	1.1	1.6
Cholesterol	1.4	2.1
Cortisol	4.1	5.6

Quality Specifications in Action   We now have a hierarchy of methods for setting quality specifications. We should use models as high up the hierarchy as possible. However, the best approach (i.e., analysis of the effect of analytical performance on any measure of clinical outcome) is difficult, because tests are used in different clinical situations, and because clinicians have difficulty describing objectively how they use test results.

In general, then, quality specifications based on biology are appropriate and readily available in the literature and on the Internet: they exist for precision, bias, total error allowable, allowable difference between two methods, drugs, fixed limits for use in PT and EQAS, and reference methodology The characteristics of precision, bias, and total error are, of course, the most important for quality planning.

Desirable quality specifications for the most commonly requested tests in clinical biochemistry, urine analytes, and hematology and hemostasis are shown in Appendix 2 and optimum and minimum specifications can be easily calculated from the data in Appendix 1.

Laboratories should know their precision (from their internal quality control program) and their bias (from comparing their results with other methods of known bias when the method was set up, or more usually, from comparison of results with values obtained in PT or EQAS). The question then arises as to what level of performance ensures that quality specifications are met. Some years ago, it was suggested that if the precision were less than one-third of the quality specification, then the method was satisfactory; this rather simplistic (but useful at the time) approach assumed that bias was not a problem.

Westgard has suggested an easy way to judge whether or not a method is satisfactory. It is called the *method decision chart* (www.westgard.com/lesson25.htm).

- First you calculate the total error allowable quality specification.
- Then you prepare a piece of graph paper with the bias on the *y*-axis and precision on the *x*-axis. The range of the *y*-axis is zero to TE$_a$ and the scale is the ratio of bias to TE$_a$ (B$_A$/TE$_a$). The range of the *x*-axis is TE$_a$/2 and the scale is the ratio of precision to TE$_a$ (CV$_A$/TE$_a$).

QUALITY SPECIFICATIONS

- Next, mark points on the *x*-axis at $TE_a/3$ and $TE_a/4$. Draw lines from TE_a on the *y*-axis to $TE_a/4$, $TE_a/3$, and $TE_a/2$ on the *x*-axis, which divides the area into 4 segments: excellent performance, good performance, marginal performance, and unsatisfactory performance.

Essentially Westgard is using various possible total error formulae (B_A + 4SD, B_A + 3SD, and B_A + 2SD) to create these lines, then plotting the actual precision and bias on the chart as ratios of TE_a.

This chart could be made rather more sophisticated and useful by including the formulae we have used in this book, in particular by extending the *x*-axis to $TE_a/1.65$ (0.6 TE_a) and connecting a line from TE_a on the *y*-axis to $TE_a/1.65$. The area to the left of the poor performance segment would then be called unacceptable performance (see Figure 2.9).

According to the model, a method with excellent performance would be more

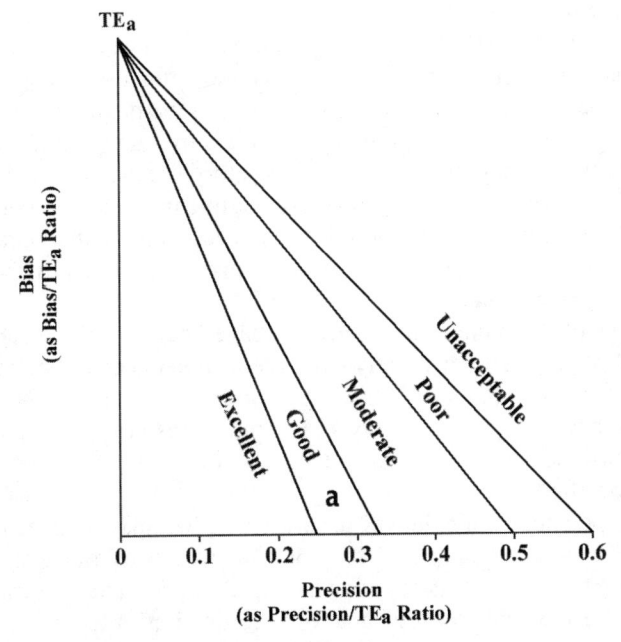

Figure 2.9 Modified Method Decision Chart

Here is an example using Westgard's model. We know (from data on biological variation) that the desirable quality specification for total error for serum albumin is 3.9%. We know that our laboratory method always has a positive bias compared to the overall mean (derived by consensus) in our EQAS of 0.4% (i.e., about 1/10 of the TE_a). Our method precision at the lower reference limit is 1.0% (i.e., about 1/4 of the TE_a). Plotting 1/10 on the *y*-axis (our bias) against 1/4 on the *x*-axis (our precision) gives point **a** on the figure. Thus, our albumin method can be described as "good." It is important to note that an increase in *either* bias *or* precision might make this method fall into an inferior quality category; this has important consequences for quality management.

than acceptable because it will be easy to manage in routine service and easy to control quality. A method with good performance would meet quality requirements and could be managed in routine service with careful quality planning. A method with marginal performance provides the desired quality when everything is working correctly but would be very difficult to manage in routine practice without a great deal of attention to all facets of the quality management cycle. A method with poor performance does not meet the quality specifications and is unacceptable for day-to-day, routine service. If the modification suggested here and shown in Figure 2.9 were used, a method with poor performance would barely meet the quality specifications and would likely be found unacceptable for day-to-day routine service. The meaning of the final segment is obvious: the method would be unacceptable.

The method decision chart could be made even more sophisticated, using what is currently in vogue: "Six Sigma Quality Management," i.e., we should achieve performance that is one-sixth of the quality specification. We have seen that mean ± 2SD encompasses 95.5% of a distribution so that, by definition 4.5% of the data will be outside the ± 2SD interval even when the method is working perfectly. If we use ± 6SD, then the number of values outside the interval is minimal—only one in 500 million! Thus, if we achieve this process standard, variations in the process will not cause the results to fail the quality specifications.

We could apply six sigma quality management by simply dividing our quality specifications by six and comparing our laboratory precision to these (assuming negligible bias). Alternatively, we could draw a line in our method decision chart from TE_a on the y-axis to $TEa/6$ on the x-axis—a method with characteristics to the left of this line would meet the demanding six sigma standard and could perhaps be described as "superb."

The basics of these interesting concepts have been recently discussed on the Internet by Westgard. The seemingly not widely appreciated but basic underlying assumption of quality planning is this: if a method is perfect, then *no* quality control is needed because quality is a given. If a method is nearly perfect, only occasional errors will be made and so little quality control will be required to guarantee the quality specified.

A further potentially simple way to assess whether analytical techniques meet quality specifications is through calculating what is termed Critical Systematic Error (abbreviated as ΔSE_c). It is easy to calculate ΔSE_c from the information we already have. We have defined our quality specification in the form of TE_a. We know our precision (CV_A) and our bias (B_A) from the sources discussed above. Then, the calculation required is—

$$\Delta SE_c = [(TE_a - B_A)/CV_A] - 1.65.$$

ΔSE_c is a good indicator of our method performance relative to quality specifications. ΔSE_c indicates, in a single statistic, the number of SD the mean can shift before exceeding the total error allowable quality specification. Therefore, ΔSE_c = 10 equates to a method where the mean can shift 10 SD before exceeding the quality specifications, whereas $\Delta SE_c = 2$ indicates the mean can only shift 2 SD. The concept of ΔSE_c combines four key parameters into a single statistic that tells us where we are relative to where we need to be.

For example, from the data on within-subject and between-subject biological variation of serum magnesium (Appendix 1), we know that CV_I is 3.6% and CV_G is 6.4%, so that $TE_a = 4.8\%$. Now, if we had bias of 0.2% and precision of 1.2%, then $\Delta SE_c = [(4.8-0.2)/1.2]-1.65 = 2.2$.

On the other hand, from data on within-subject and between-subject biological variation of serum phosphate, we know that CV_I is 8.5% and CV_G is 9.4%, so that $TE_a = 10.2\%$. If we had the identical performance characteristics of bias of 0.2% and precision of 1.2%, then $\Delta SE_c = [(10.2-0.2)/1.2]-1.65 = 6.7$.

A method with high ΔSE_c (> 3.0) will easily meet requirements for quality and could be managed in routine service with simple quality control. A method with intermediate ΔSE_c (2.0–3.0) provides the desired quality when everything is working correctly but would be more difficult to manage in terms of internal quality control requiring possibly more control samples per run or more stringent rules for acceptance or rejection. A method with low ΔSE_c (<2.0) will very difficult to manage in routine practice without a great deal of attention to all facets of the quality management and may well be unacceptable for routine service.

If the laboratory cannot achieve quality specifications, it can take one of several approaches.

- Not worry and continue to use the current method and traditional statistical quality control approaches. This is a poor strategy because the laboratory is continuing to monitor what is intrinsically unsatisfactory performance.
- Cease performing the test because it was not of sufficient quality. However, this would eliminate many useful tests from the repertoires of a number of laboratories.

Choosing one of the more positive approaches (listed below) is recommended.

- Consider applying another set of quality specifications from lower in the hierarchy, especially those that would allow the laboratory to retain accreditation through satisfactory participation in PT or meet the usually less demanding standards of the fixed limits in EQAS.
- Institute quality improvement through application of techniques for quality laboratory practice.
- Take steps to improve the methodology itself.
- Investigate alternative methodology and technology that will allow quality specifications to be met.

If the laboratory is in the happy situation of surpassing even minimum quality specifications, then the laboratory should not allow performance to deteriorate, but should apply quality planning to guarantee that the quality specifications are met while saving internal quality control resources.

The application of quality specifications in quality planning is rather too complex to be described in detail in this book. There are a number of different approaches. The principles of quality planning can be delineated as follows.

- Document quality specifications for precision, bias, and total error allowable.
- Measure precision from internal quality control and measure bias from comparison of methods, PT or EQAS, or peer comparison programs for QC samples.
- Compare these to decide objectively the number of controls per run and the control rules to be applied in order to guarantee the quality specified. This process is described in a number of sources, including the very detailed and comprehensive site on the Internet (www.westgard.com) and in detail in a text by Zoe Brooks (*Performance-Driven Quality Control*, Washington, DC: AACC Press, 2001).

The main point to note is that all approaches require clearly defined quality specifications at the outset.

SUMMARY

The following key points were covered in this chapter.

- Different individuals may have different views on the desirable characteristics of laboratory tests. The level of performance required to facilitate clinical decision-making is the quality specification.
- All methods can be fully described in terms of their performance characteristics: practicability characteristics and reliability characteristics.
- Quality specifications are required for many purposes in the commissioning of new analytical systems and in daily quality management, particularly for precision, bias, and total error allowable.
- Total error can be calculated by simple linear addition of precision and bias: various formulae exist.
- Many difficulties with the setting of numerical quality specifications for precision and bias have been documented.
- Available models have been arranged in a hierarchy in a global strategy for setting quality specifications in laboratory medicine. This hierarchy has been accepted as consensus by expert professionals.
- All models for setting quality specifications have advantages and disadvantages.
- Ideally, quality specifications are set by analyzing the effect of analytical performance on clinical outcomes. However, this is difficult because few tests are used in specific clinical situations and it is difficult to determine how clinicians use test results.
- Quality specifications can be generated using biological variation, probabil-

ity, and the results of vignette studies used to seek the views of clinicians. In general, these studies have not been well designed or executed in the past but seem to have potential.
- National and international expert groups and individual experts have proposed quality specifications: many of these use biological variation as their basis.
- Quality specifications can be derived from the fixed limits laid down in PT and EQAS but these are much influenced by the state of the art.
- If nothing else is available, the state of the art from PT, EQAS, or publications can be used to derive quality specifications, but these bear little relationship to clinical needs.
- Models higher in the hierarchy are preferred to those lower in the hierarchy.
- Variation must be added as variance, i.e., squares of SD or CV.
- Precision affects test result variability but not in a linear relationship.
- Inferior precision makes changes seen in serial results from an individual of less significance, and leads to wide reference intervals.
- Quality specifications for precision are generally best based on biology. It is widely accepted that precision should be less than one-half the within-subject biological variation.
- Biological variation can also be used to generate minimum quality specifications for analytes for which it is difficult to meet the desirable standards, and can be used to generate optimum quality specifications for analytes for which it is easy to meet the desirable standards.
- The utility of reference values is much influenced by bias.
- Quality specifications for bias are generally best based on biology. Bias should be less than one quarter of the group (within-subject plus between-subject) biological variation.
- Biological variation can also be used to generate minimum quality specifications for analytes for which it is difficult to meet desirable standards, and can be used to generate optimum quality specifications for analytes for which it is easy to meet desirable standards.
- Quality specifications for total error allowable for use in quality planning models can be derived by linear addition of precision and bias specifications.
- Quality specification for other purposes can also be derived from biological variation.
- Data are easily available for desirable quality specifications for precision, bias, and total error allowable for more than 300 analytes.
- When quality specifications based on biology are not met, other quality specifications in the hierarchy might be applied but quality improvement should be instituted.
- The acceptability of the performance characteristics can be judged using a method decision chart or through calculating critical systematic error.
- When quality specifications based on biology are easily achieved, institution of quality planning will save resources.

FURTHER READING

1. Brooks Z. Performance-driven quality control. Washington, DC: AACC Press, 2001.
2. C24–A2. Statistical Quality Control for Quantitative Measurements: Principles and Definitions; Approved Guidelines, 2nd Edition. Wayne, PA (USA): National Committee for Clinical Laboratory Standards, 1999.
3. Fraser CG, Hyltoft Petersen P, Libeer JC, Ricos C. Proposals for setting generally applicable quality goals solely based on biology. Ann Clin Biochem 1997;34:8–12.
4. Fraser CG, Hyltoft Petersen P, Ricos C, Haeckel R. Proposed quality specifications for the imprecision and inaccuracy of analytical systems in clinical chemistry. Eur J Clin Chem Clin Biochem 1992;30:311–317.
5. Fraser CG, Hyltoft Petersen P. Analytical performance characteristics should be judged against objective quality specifications. Clin Chem 1999;45:321–323.
6. Fraser CG, Hyltoft Petersen P. The importance of imprecision. Ann Clin Biochem 1991; 28:207–211.
7. Fraser CG. Judgment on analytical quality requirements from published clinical vignette studies is flawed. Clin Chem Lab Med 1999;37:167–168.
8. Gowans EMS, Hyltoft Petersen P, Blaabjerg O, et al. Analytical goals for the acceptance of common reference intervals for laboratories throughout a geographical area. Scand J Clin Lab Invest 1988;48:757–764.
9. Hyltoft Petersen P, deVerdier CH, Groth T, Fraser CG, et al. The influence of analytical bias on diagnostic misclassifications. Clin Chim Acta 1997;260:189–206.
10. Hyltoft Petersen P, Fraser CG, Kallner A, Kenny D, eds. Strategies to set global analytical quality specifications in laboratory medicine. Scan J Clin Lab Invest 1999;57:475–585.
11. Skeie S, Thue G, Sandberg S. Patient-derived quality specifications for instruments used in self-monitoring of blood glucose. Clin Chem 2001;47:67–73.
12. Thienpont L, Franzini C, Kratochvila J, et al. Analytical quality specifications for reference methods and operating specifications for networks of reference laboratories. Eur J Clin Chem Clin Biochem 1995;33:949–957.
13. U.S. Dept. of Health and Human Services. Medicare, Medicaid, and CLIA Programs: Regulations implementing the Clinical Laboratory Improvement Amendments of 1988 (CLIA). Final rule. Fed Regist 1992;57:7002–7186.
14. Westgard JO. Six sigma management and desirable laboratory precision. Available online at www.westgard.com/essay35.htm.

Chapter 3
Changes in Serial Results

Clinical laboratory test results are used for many purposes. We use them in teaching and training, research and development, and in four rather different clinical situations, namely, diagnosis, case finding, screening, and monitoring.

Most test results are used to monitor people's health over the short, medium, or long term. As part of a quality service, we must help clinicians objectively interpret the numbers we report from our laboratories. We should provide objective guidance in clinical monitoring about the significance of changes in serial results.

CALCULATING REFERENCE CHANGE VALUES AND PROBABILITY

Monitoring with Clinical Laboratory Test Results   The 63-year-old male with the "high risk" serum cholesterol concentration we studied earlier actually presented to his general practitioner complaining of headache, dizziness, irritability, fatigue, and insomnia. These are rather non-specific complaints but could be adequately explained by the practitioner's finding of high blood pressure (hypertension). The practitioner then requested an appropriate panel of laboratory tests including investigation of renal function, liver function, lipids, and thyroid function. The results are shown as the 1st Result in Table 3.1.

The general practitioner was concerned about the less-than-ideal serum cholesterol concentration. His treatment included advice about lifestyle modification including the need to stop smoking, reduce weight, reduce alcohol intake, be physically active, and modify diet. After one month, the tests were requested again. The results are shown in the last column of Table 3.1 (2nd Result).

All the numbers have changed. We can see that the cholesterol concentration is now 5.82 mmol/L and this is a lower number than 6.60 mmol/L. We already know that there are many sources of variation inherent in any test result but the clinical question posed here is, *"Are any of the changes significant; in particular, has the serum cholesterol concentration fallen significantly simply due to lifestyle modification?"*.

We could turn this question into a scientific one. We know that all analytes assayed in the clinical laboratory have inherent variation. The question could be posed as this: *"Is the change in serum cholesterol larger than we would expect from these intrinsic sources of variation?,* or in more general terms, *"How do we interpret serial results obtained on individual patients?"*

Interpretating Change against Fixed Criteria   There are many aids to clinical interpretation of numerical laboratory results. For example, test results can be compared to

Table 3.1 Test Results in a 63-Year-Old Male with Hypertension

Analyte	1st Result	Units	Reference Interval	2nd Result
Sodium	144	mmol/L	135–147	142
Potassium	4.6	mmol/L	3.5–5.0	4.4
Urea	5.2	mmol/L	3.3–6.6	5.4
Creatinine	102	µmol/L	64–120	97
ALT	18	U/L	13–43	14
Bilirubin	10	µmol/L	0–17	12
Alk Phos	89	U/L	30–105	81
GGT	35	U/L	11–82	39
Calcium	2.49	mmol/L	2.10–2.55	2.43
Albumin	38	g/L	35–50	42
Cholesterol	6.60	mmol/L	Ideal < 5.00	5.82
Triglycerides	1.08	mmol/L	Up to 2.30	1.32
TSH	2.03	mU/L	0.4–4.0	2.19
Urate	0.28	mmol/L	0.24–0.46	0.32

- population-based reference values (we will explore the advantages and disadvantages of using conventional population-based reference values in clinical monitoring in detail in Chapter 4),
- locally agreed protocols for clinical action,
- values proposed by expert individuals, groups, or committees,
- values based on risk, and
- multiples of the upper reference limit.

The last four interpretative aids listed above, irrespective of how they are developed, can be described as "clinical fixed limits"; sometimes these are called cut-off points. These do have advantages in that that they are in some way related to clinical outcome and are usually based on either good evidence or the views of "experts." However, they have disadvantages when used to monitor patients in the clinical setting.

Take the case of the 63-year-old-man with the elevated cholesterol. Suppose he receives a statin drug to lower his serum cholesterol concentration. Note that statins should be used with caution in those with a history of liver disease or with a high alcohol intake. Moreover, it is recommended that liver function tests be requested before treatment and regularly thereafter. Treatment should be discontinued if "serum transaminase rises to 3 times the limit of the reference range" according to the British National Formulary—this is a "fixed limit" set using another "fixed limit."

Results for the profile of liver function tests are as shown in Table 3.2.

Figure 3.1 graphs the serum ALT and AST activity changes with time. The reference interval for both ALT and for AST for men of this age is 13–43 U/L. Three times the upper reference limit is therefore 129 U/L.

Table 3.2 and Figure 3.1 demonstrate that, as expected, all analytes change with time. For serum albumin, alkaline phosphatase, and bilirubin, the changes ap-

CHANGES IN SERIAL RESULTS

Table 3.2 Results of Some Liver Function Tests

Time	Albumin (g/L)	Alk Phos (U/L)	Bilirubin (μmol/L)	ALT (U/L)	AST (U/L)
Pre-dose 1	38	89	10	18	16
Pre-dose 2	42	81	12	14	18
After 2 months	40	86	9	27	24
After 4 months	41	94	13	35	40
After 6 months	39	85	8	45	54
After 8 months	42	91	14	58	69

pear to be random. This, of course, is expected. Changes also occur for both serum ALT and AST activities. These changes do not appear to be random. Both enzyme activities appear to rise. This is unexpected. Both activities become higher than the appropriate age and sex matched upper reference limit. However, they do not exceed the clinically set criterion for stopping the therapy of three times the upper reference limit.

Should the clinician worry about these unusual results? Should laboratories draw attention to them? In this Chapter, we explore whether there are more objective ways of looking at serial results in an individual other than fixed limits, irrespective of whether they are population-based reference limits or clinically set limits which, after all, are very often empirical.

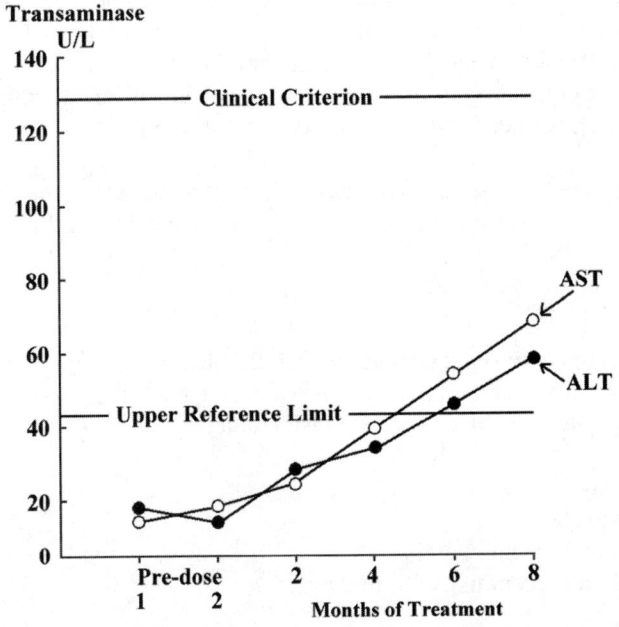

Figure 3.1 Changes in Serum ALT and AST Activities with Time

Calculating Total Variation We know that all analytes assayed in the clinical laboratory have inherent variation due to pre-analytical variation, analytical variation, and within-subject biological variation. These variations are all random and can be considered Gaussian. The dispersion of a Gaussian distribution can be described in terms of standard deviation (SD). Sources of variation are additive and, when adding variation, we must add the SD as their squares.

If

- pre-analytical variation is called SD_P,
- analytical variation is called SD_A,
- and within-subject biological variation is called SD_I,

then the total variation (SD_T) can be calculated as follows:

$$SD_T^2 = SD_P^2 + SD_A^2 + SD_I^2 \text{ or } SD_T = (SD_P^2 + SD_A^2 + SD_I^2)^{1/2}.$$

It is often advantageous in laboratory medicine to do calculations such as additions of variation using the CV calculated as (SD/mean) * 100. In this case, we can add variations as CV because the means of the components are identical. The calculation of total variation becomes

$$CV_T^2 = CV_P^2 + CV_A^2 + CV_I^2 \text{ or } CV_T = (CV_P^2 + CV_A^2 + CV_I^2)^{1/2}.$$

If we want to state with confidence that an individual's serial results have changed, then the difference in results must exceed that which can be explained by the inherent variation due to these three factors. To make small changes more significant from a clinical point of view, i.e., reduce the "noise" to make the clinical "signal" clearer, we need to reduce sources of variation.

Minimizing Pre-Analytical Variation Total variation is the sum of the component variations. For laboratory test results, we have calculated that

$$CV_T = (CV_P^2 + CV_A^2 + CV_I^2)^{1/2}.$$

We know that to make small changes in an individual more significant, we need to reduce sources of variation. As we know from Chapter 1, the possibilities for variation in the pre-analytical phase are considerable:

- time of day,
- fed/fasting state,
- previous and recent exercise,
- previous and recent use of stimulants,
- posture,
- source of sample,

- anticoagulant or preservative or stabilizer type,
- tourniquet application period,
- transport time and temperature,
- centrifugation time and force, and
- storage conditions.

If we

- standardize the conditions of patient preparation including time of day and posture in particular,
- adopt standard procedures for phlebotomy including types of sample collected and containers used, and tourniquet application time, and
- adhere to standard types of sample transport, handling, and centrifugation,

then we can minimize pre-analytical variation. If pre-analytical variation is minimized, then since CV_P is irrelevant, our total variation equation is simplified to this:

$$CV_T = (CV_A^2 + CV_I^2)^{1/2}.$$

This formula holds when, as is current laboratory practice, one sample is taken and this is analyzed once only. Analyzing more than one sample more than once reduces CV_T, but this is rarely done in practice. Duplicate analyses are still quite often performed in procedures with manual steps, such as radioimmunoassays, but the aim here is really to catch "blunders"—major mistakes such as missing addition of a crucial reagent—rather than reducing random variation. If we take more than one sample, or if we perform analysis on one or more samples in replicate, different, slightly more complex formulae must be used. The relevant CV then is divided by the square root of the number of times the sample is collected or the assay performed—so the general formula is

$$CV_T = (CV_A^2/n_1 + CV_I^2/n_2)^{1/2},$$

where n_1 is the number of analytical replicates and n_2 is the number of simultaneous samples taken. Interestingly, doing replicate analyses does not reduce the variation in a linear manner—doing the assay twice makes the variation 70% that of doing it once ($1/2^{1/2} * 100$); doing the assay three times makes the variation 58%; and four times, 50% of the singleton value.

Reference Change Values Assume we have minimized pre-analytical variation through good training, good laboratory practice, and adherence to good written standard operating procedures. Then, the total variation for each laboratory result (single sample, single analysis) is

$$CV_T = (CV_A^2 + CV_I^2)^{1/2}.$$

This variation is random and therefore Gaussianly distributed. It is therefore easy to derive the range in which the values will lie with a certain probability.

To review, we know that, for a Gaussian distribution:

- The found value will lie within the range value ± 1 CV with 68.3% probability.
- The found value will lie within the range value ± 2 CV with 95.5% probability.
- The found value will lie within the range value ± 3 CV with 99.7% probability.

The multipliers 1, 2, and 3 are called the *Standard Normal Deviates* and are also called *Z-scores*. Thus, the found value for any analyte lies within

$$\pm Z * (CV_A^2 + CV_I^2)^{1/2}$$

with a probability appropriate to the Z-score (for more information about Z-scores, consult standard statistical tables).

Note that we include ± when calculating the ranges shown above. This makes Z scores bi-directional. Of course, we can easily work out the range for +1 CV or +2 CV or +3 CV and unidirectional Z-scores can also be found in standard statistical tables. Knowing this is important when we consider serial results. A simplified listing of Z-scores for common probabilities for bidirectional and unidirectional situations is shown in Table 3.3 (note the non-linear relationship between probability and value of Z).

When we consider differences in two serial results for an analyte, each result has a variation equal to

$$Z * (CV_A^2 + CV_I^2)^{1/2}.$$

Table 3.3 Z-scores and Probability

Probability (%)	Unidirectional Z-score	Bidirectional Z-score
99	2.33	2.58
98	2.05	2.33
97	1.88	2.17
96	1.75	2.05
95	1.65	1.96
90	1.28	1.65
85	1.04	1.44
80	0.84	1.28
75	0.68	1.15
70	0.52	1.04
60	0.25	0.84

CHANGES IN SERIAL RESULTS

We have two results—one before and one after—so, since total variation is the sum of the individual variations (added, of course, as variances), then

total variation = [(variation of first test)^2 + (variation of second test)^2]$^{1/2}$

which is

total variation = $\{[Z*(CV_A^2 + CV_I^2)^{1/2}]^2 + [Z*(CV_A^2 + CV_I^2)^{1/2}]^2\}^{1/2}$

which simplifies to

total variation = $2^{1/2} * Z * (CV_A^2 + CV_I^2)^{1/2}$.

Thus, for serial results to be significantly different, the difference in numerical results must be greater than the combined variation inherent in the two results. This value is traditionally called the *critical difference* but is better called the *reference change value (RCV)* to keep nomenclature consistent with international recommendations (discussed in more detail in Chapter 4). It is easy to calculate as

RCV = $2^{1/2} * Z * (CV_A^2 + CV_I^2)^{1/2}$.

Calculation of RCV We can calculate reference change values using this formula:

RCV = $2^{1/2} * Z * (CV_A^2 + CV_I^2)^{1/2}$.

Z-scores are easily available and the most commonly used are shown in Table 3.3. We may *want* to ask the question whether we want bidirectional or unidirectional RCV. The question we *must* ask is this: "Are we interested in whether the result has changed (risen or fallen, i.e., bi-directional) which is, as expected, a bigger window than that obtained by asking whether the result has increased or decreased (unidirectional)?".

In usual practice, we want to know about change in general and so we apply bidirectional Z-scores. Few people have ever used anything but the bidirectional approach (and this is what we use in the examples in this book). In addition, 95% probability is conventionally regarded as *significant* and 99% probability is conventionally regarded as *highly significant*. In consequence, generally 1.96 and 2.58 are the appropriate Z-scores to use.

CV_A is the analytical variation (analytical precision). We should understand that every laboratory should know its own precision for every analyte assayed through its internal quality control program. Ideally, then, in the formula we should use the CV_A appropriate to the mean value between the first and second results. However, internal quality control is usually done only at a few levels of analyte. These levels, then, should be at critical decision-making points because they are where we require best performance. Thus, use the CV_A from the quality control at the appropriate clinical decision-making level.

CV_I is the within-subject biological variation (easily obtained from literature cited in Chapter 1 and, for the most common analytes, in Appendix 1). We have seen that the estimates of within-subject variation are rather robust and are transferable across time and geography so that each laboratory can use values cited in the literature to create its own relevant RCV.

Thus it is easy to calculate RCV for significant and highly significant changes:

- Use $2^{1/2}$ because we have 2 samples.
- Use 1.96 and 2.58 as the Z-scores for *significant* and *highly significant* RCV.
- Use CV_A from your own internal QC program at the appropriate decision-making level.
- Use CV_I from the most recent database available.

Recall that our 63-year-old hypertensive man had serum cholesterol concentration of 6.60 mmol/L which, on lifestyle modification, "changed" to 5.82 mmol/L. But, the real question is, "Has the cholesterol concentration changed significantly?".

We can calculate the answer as follows:

First value   = 6.60 mmol/L
Second value   = 5.82 mmol/L
Change   = 6.60 − 5.82 = 0.78 mmol/L
equivalent to (0.78/6.60) * 100 = 11.8%

Note that change is the percentage of difference from the first value:

$$RCV = 2^{1/2} * Z * (CV_A^2 + CV_I^2)^{1/2}$$
$$2^{1/2} = 1.414$$

and

Z = 1.96 for a significant change—that is, a change with 95% probability (p<0.05)
Z = 2.58 for a highly significant change—that is, a change with 99% probability (p<0.01)

CV_A is taken from the internal quality control run in the laboratory. Over the time period of these cholesterol assays, we ran 201 assays of our control at 6.91 mmol/L and achieved the SD of 0.11 mmol/L. Therefore,

CV_A is (0.11/6.91) * 100 = 1.6%.

CHANGES IN SERIAL RESULTS

CV_I is taken from the latest published database and is 6.0%. Thus, for a significant change

$$RCV = 2^{1/2} * Z * (CV_A^2 + CV_I^2)^{1/2}$$
$$RCV = 1.414 * 1.96 * (1.6^2 + 6.0^2)^{1/2} = 17.2\%$$

And for a highly significant change

$$RCV = 2^{1/2} * Z * (CV_A^2 + CV_I^2)^{1/2}$$
$$RCV = 1.414 * 2.58 * (1.6^2 + 6.0^2)^{1/2} = 22.6\%$$

Thus, while the two results are different, and it looks as if the second number is lower than the first, the serum cholesterol has not changed significantly ($p<0.05$) or highly significantly ($p<0.01$)—at least from an objective statistical point of view.

We can "see" that this finding (that a serum cholesterol of 6.60 mmol/L) is not significantly or highly significantly different from a serum cholesterol of 5.82 mmol/L) by considering the real dispersions of the numerical test results. We have seen already that the 95% dispersion is ± 1.96 SD and the 99% dispersion is ± 2.58 SD. Thus, since this dispersion is made up of $(SD_A^2 + SD_I^2)^{1/2}$ and SD can easily be calculated from CV as SD = (CV * value)/100, we can determine that:

- The 99% dispersion of a serum cholesterol of 6.60 mmol/L is 5.54–7.66 mmol/L.
- The 95% dispersion of a serum cholesterol of 6.60 mmol/L is 5.80–7.40 mmol/L.
- The 95% dispersion of a serum cholesterol of 5.82 mmol/L is 5.11–6.53 mmol/L.
- The 99% dispersion of a serum cholesterol of 5.82 mmol/L is 4.89–6.75 mmol/L.

The dispersions overlap considerably (see Figure 3.2) due to inherent precision and within-subject biological variation. The results are not significantly or highly significantly different.

The Problem of "Random" Bias Random changes in bias due, for example, to re-calibration and changes of reagent lots, do contribute to test result variation over time. With modern analytical systems, these changes in bias are sometimes bigger than the inherent precision. Of course, we should eliminate these where and when possible, and certainly minimize them by careful attention to quality management at re-calibration.

If we can quantify change in bias between two results—and let us call this

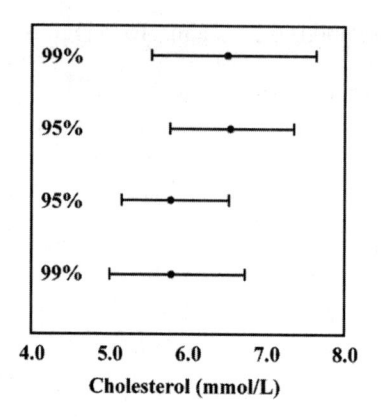

Figure 3.2 95% and 99% Dispersions of Serum Cholesterol Results of 6.60 and 5.82 mmol/L

Test results together with 95% and 99% dispersions. Results are not significantly or highly significantly different.

ΔB—then we could make our formula more sophisticated by linearly adding bias and precision (just like total error). The RCV would then become

$$RCV = \Delta B + 2^{1/2} * Z * (CV_A^2 + CV_I^2)^{1/2}.$$

However, it is really difficult to quantify these changes in bias with time numerically, and, over the longer term, these changes, which can be termed "random bias," are most often included in the precision calculations that use data from analyses of internal quality control samples.

RCV for Liver Function Tests We can use the RCV to assess whether test results on statin have changed from pre-dose values to the values at 2 months post-dose in our patient exactly as we have done for cholesterol in the example previously.

We can calculate the RCV using the formula

$$RCV = 2^{1/2} * Z * (CV_A^2 + CV_I^2)^{1/2}$$

and, for liver function tests, create a "spreadsheet," which may make the calculation easier, and fill in the items required for the calculation as shown in Table 3.4.

Results on our 63-year-old male on statin were as shown earlier in Table 3.2. We can now calculate the significance of the change as shown in Table 3.5.

The changes for serum ALT and AST activities are highly significant ($p < 0.01$) and significant ($p < 0.05$) respectively. The values at 2 months post-dose have indeed changed from the pre-dose activities. The values post-dose are *not* above the reference limit. They are *not* above the clinical criterion for cessation of therapy. Thus, using RCV as an interpretative tool for assessment of serial results, we have provided early warning of an upward change which is confirmed by the values found at 4, 6, and 8 months.

CHANGES IN SERIAL RESULTS

Table 3.4 Data for Spreadsheet to Calculate RCV for Liver Function Tests

Analyte	$2^{1/2}$	Z for 95%	Z for 99%	CV_A (from internal QC)	CV_I (from database)	RCV: 95%	RCV: 99%
Albumin	1.414	1.96	2.58	0.8	3.1	8.6	11.7
Alk Phos	1.414	1.96	2.58	1.4	6.4	18.1	23.9
Bilirubin	1.414	1.96	2.58	1.0	25.6	70.9	93.4
ALT	1.414	1.96	2.58	0.9	24.3	67.3	88.7
AST	1.414	1.96	2.58	1.1	11.9	33.2	43.8

Probability that a Change Is Significant   Our 63-year-old man had serum cholesterol concentration of 6.60 mmol/L which, on lifestyle modification, "changed" to 5.82 mmol/L. After investigating whether the change was significant, we concluded that the cholesterol concentration had neither changed significantly ($p < 0.05$) nor highly significantly ($p < 0.01$).

The numbers do look different. Then what is the probability that the particular change is significant?

We calculate reference change values (RCV) using the formula

$$RCV = 2^{1/2} * Z * (CV_A^2 + CV_I^2)^{1/2}.$$

Now, if we know the change in serial results, and if we want to know the probability, we simply rearrange the formula to make Z the unknown:

$$Z = \text{Change}/[2^{1/2} * (CV_A^2 + CV_I^2)^{1/2}].$$

For our 63-year-old man with lifestyle modification, for cholesterol,

Table 3.5 Calculation of the Significance of Change in Selected Liver Function Tests

Time	Albumin (g/L)	Alk Phos U/L	Bilirubin (µmol/L)	ALT (U/L)	AST (U/L)
Pre-dose 2	42	81	12	14	18
After 2 months	40	86	9	27	24
Change (units)	2	5	3	13	6
Change (%)	4.8	6.2	25.0	92.9	33.3
Significance	NS	NS	NS	Highly significant	Significant

1st value   = 6.60 mmol/L
2nd value   = 5.82 mmol/L
Change   = 6.60 − 5.82 = 0.78 mmol/L
which is equivalent to (0.78/6.60) ∗ 100 = 11.8%

Now

$2^{1/2} = 1.414$

and, as we have seen, CV_A is found from the internal quality control run in the laboratory and was 1.6%, and CV_I is found from the latest published database and is 6.0%.

Thus $Z = 11.8 / [2^{1/2} * (1.6^2 + 6.0^2)^{1/2}] = 1.35$.

and, looking at the statistical tables, we find that, for that value of Z, the probability that this change is significant is actually quite high: 82%.

We should recognize that, while statisticians and laboratory staff traditionally like to use 95% (p<0.05) and 99% (p<0.01) levels of probability, and call these *significant* and *highly significant*, clinical medicine uses a very wide range of probabilities in real practice.

For example, we might want to be *certain* that a change in a serum marker concentration had occurred before instituting unpleasant therapy or even further costly investigations. Other terms with very high probability are *convince, assure, cause,* and *prove*. These might be regarded as 99% probability.

On the other hand, we might be happy that it was *hinted* that the serum cholesterol had changed in our 63-year-old man on lifestyle modification alone and this might prompt us to suggest he continues with this approach, eliminating both the costs and risks of drug therapy. Other terms suggesting low probability are *indicate* and *draw attention to*. These might be regarded as 80% probability.

Intermediate terms such as *prompt, suggest, advise,* and *propose* might be regarded as being equivalent to the significant level of 95% probability.

Other terms such as *likely* and *lead one to believe* might be considered to have probability of 90%.

The problem is that different people use different semantics in different settings in different situations for different levels of probability. In consequence, we should not assume that just because a change in results is not statistically significant, that the change is clinically unimportant. On the other hand, a change in results may be highly significant statistically but not warrant any clinical action.

QUALITY MONITORING OF PATIENTS IN PRACTICE

We use RCV to determine whether changes in an individual's serial results are significant, and we can rearrange the formula to determine the probability of sig-

CHANGES IN SERIAL RESULTS

nificance. These are easy to calculate from readily available data. Laboratories need to put these simple tools into everyday practice. Therefore let us examine the factors that affect RCV and the strategies that are available to us to ensure that consumers of our services actually use these aids to interpretation.

Precision and RCV Recall the formula for calculating reference change value:

$$RCV = 2^{1/2} * Z * (CV_A^2 + CV_I^2)^{1/2}.$$

We know that the factor $2^{1/2}$ is a constant (1.414); Z is a constant, depending on the probability; and CV_I can be assumed to be a constant—at least in apparently healthy people and in those with chronic but stable disease. Thus, if we want to decrease RCV, i.e., we want smaller changes in serial results to be significant, the only option that we have is to decrease CV_A. We must reduce the analytical precision. By doing this, we make the "noise" less important in relation to the "signal."

Recall that the serum cholesterol in our 63-year-old man was 6.60 mmol//L on the first analysis and was 5.82 mmol/L on the second; RCV (for 95 % probability) for cholesterol was 17.2% with the precision achieved in the laboratory. Table 3.6 shows the effect of precision on the RCV for cholesterol. This is easy to calculate since, for cholesterol, $2^{1/2}$ is 1.414, Z for $p<0.05$ is 1.96, and CV_A is 6.0%, and so

$$RCV = 2^{1/2} * Z * (CV_A^2 + CV_I^2)^{1/2} = 2.77 * (CV_A^2 + 36)^{1/2}.$$

We can draw the following conclusions.

- Poor precision increases RCV.
- Good precision decreases RCV.

Table 3.6 Precision and Reference Change Values

CV_A	RCV
0	16.6
1.0	16.8
2.0	17.5
3.0	18.6
4.0	20.0
5.0	21.6
6.0	23.5
7.0	25.5
8.0	27.7
9.0	30.0
10.0	32.3

- RCV is NOT linearly related to precision.
- RCV is limited by CV_I and even at negligible precision, RCV may be quite large.
- Analytes with large inherent CV_I will have large RCV.

The probability of change in cholesterol of 11.8% achieved in our 63-year-old man was actually quite high at 82% (p<0.18). We calculated this probability from a simple rearrangement of the RCV formula. Again, probability of a change being significant depends on precision.

For this example, let us assume a change in cholesterol of 20% and use the formula

$$Z = \text{Change}/[2^{1/2} * (CV_A^2 + CV_I^2)^{1/2}].$$

The factor $2^{1/2}$ is a constant (1.414), CV_I is constant and equal to 6.0%, and the change is constant at 20%. It is Z that is variable depending on the precision, so the probability is variable. Table 3.7 shows the effect of precision on the probability that a change of 20% is significant calculated from $Z = 20/[1.414 * (CV_A^2 + 36)^{1/2}]$.

We can draw the following conclusions.

- Poor precision leads to lower probability that change is significant.
- Good precision leads to higher probability that change is significant.
- The probability of change is NOT directly and linearly related to precision.
- Probability of change is limited by CV_I and even at negligible precision probability will not be 100%.
- The probability that small change is significant will be low for analytes with large inherent CV_I.

Table 3.7 Precision and Probability That a 20% Change in Serum Cholesterol is Significant

CV_A	Z	Probability
0	2.36	98.2
1.0	2.33	98.0
2.0	2.24	97.5
3.0	2.11	96.5
4.0	1.96	95.0
5.0	1.81	93.0
6.0	1.66	90.3
7.0	1.53	87.4
8.0	1.41	84.1
9.0	1.31	81.0
10.0	1.21	77.4

CHANGES IN SERIAL RESULTS

Potential Problems with RCV   We have used the mean of the within-subject biological variation found in apparently healthy people to calculate RCV. In general, we use the data already available in the literature.

This approach has some theoretical disadvantages. First, much monitoring is conducted on hospitalized patients. These acutely ill people are often receiving intensive therapy. It is hardly surprising then, that the few studies on acutely ill people have shown that within-subject biological variation is larger in acutely ill people than in healthy people.

It has also been suggested that within-subject variation of patients in different hospital units varies depending on how ill they are and on how intensively they are treated. An example is shown in Figure 3.3. The graph shows the expected proportion of patients as the cumulative probability—i.e., the fraction of the total number—in which a change in serum sodium would be significant with 95% probability.

Samples drawn daily for at least four days showed that serum sodium in surgical intensive care patients changes more than in obstetrics and gynecology patients or in orthopedic surgery patients.

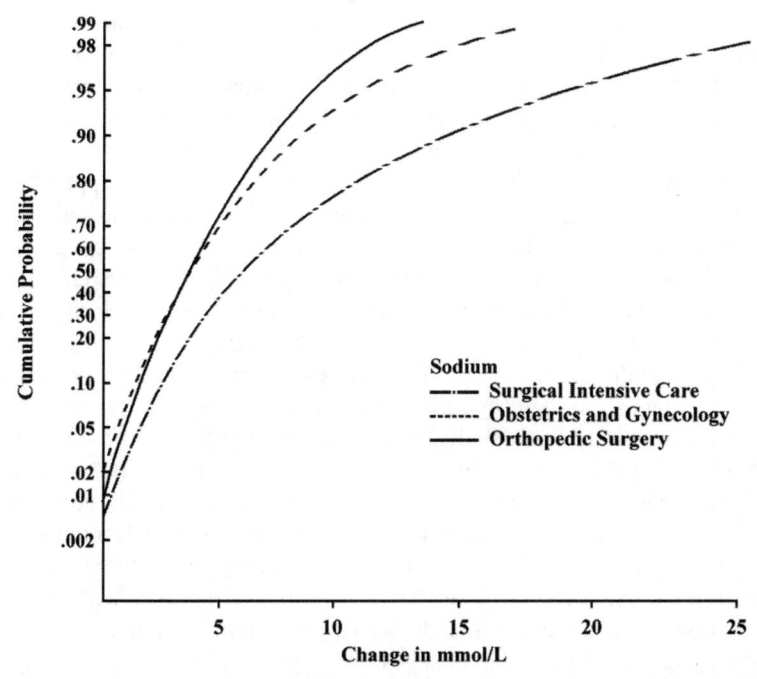

Figure 3.3   Expected Proportion of Patients (as Fraction of Total) from Different Clinical Units in which a Serum Sodium Change would be Significant with 95% Probability. (*Adapted with permission from Boyd JC, Harris EK. Utility of reference changes for the monitoring on inpatient data. In: Zinder, O, ed. Optimal use of the clinical laboratory. Basel, Switzerland, Karger, 1986.*)

In consequence, use of the healthy population RCV will lead to labeling many hospitalized patients' test results as having changed significantly. These changes could be called "false positives" since the real RCV in these patients would be higher than in the healthy. However, it is commonly considered better to draw attention to a change and then decide that it is unimportant clinically than to use population-based reference values or fixed limits as the criteria for change—and miss interesting changes in results.

A second disadvantage to using the RCV published in the literature is the general assumption that the values found in the subjects studied are a Gaussian distribution. The model assumes the variation is truly random. This means that we assume no correlation between successive results.

This assumption seems reasonable when we perform tests at medium- to long-term time intervals between samples. However, when we perform tests a number of times in a single day or on a daily basis—which again happens in hospitals, especially in those for the acutely ill—it is likely (and there is quantitative evidence to support the thesis) that serial correlation will exist. Estimates of within-subject biological variation over short time periods, even in healthy people, are smaller than longer-term estimates.

In addition, some studies have particularly looked at the correlation between successive results and shown that what is termed auto-correlation does exist. Auto-correlation makes the CV_I—and therefore the RCV—smaller. Using the "healthy" RCV will lead to less frequently labeling these patients' tests as having changed significantly. These changes could be called "false negatives."

These two effects will tend to balance each other out. Thus, calculating RCV in the easy way detailed in this book could be considered widely applicable.

A third possible problem with calculating RCV is that we usually use the mean within-subject biological variation to do the calculation. But people do have different CV_I—some are small and others are large. We can investigate this formally using a variety of statistical procedures to look at the "heterogeneity" of within-subject biological variation. Thankfully, it seems that the observed variation in CV_I for many commonly assayed analytes is only what we would expect: that the variations would simply reflect statistical randomness.

However, we could calculate RCV based upon CV_I other than the mean value: we could use the 90th percentile, for example, as some authors have recommended. This leads to larger RCV, of course. Thus, if we use this larger RCV, then we will reduce the number of "false positives" and any changes labeled as significant will likely be very important. Using the usual mean (and therefore smaller) RCV, we tend to get "false positives." However, we really do consider it better to draw attention to a change and then decide that it is unimportant than to use these larger RCV and miss interesting changes.

RCV in Practice It is easy to calculate RCV from readily available data. We can put these into practice in a number of ways.

User Guides or Laboratory Handbooks. National and international guidelines provide information on what such guides and books should contain. For laborato-

CHANGES IN SERIAL RESULTS

ries without sophisticated Laboratory Information Management Systems (LIMS), creating user guides may be the easiest way to provide information on the significance of changes in an individual's serial results. It would be easy for every laboratory to generate a simple list of RCV at 95% and 99% probabilities.

RCV are generally calculated as percentage of change. One of the difficulties is that users are given test results as simple numbers with appropriate units. Users find it difficult to calculate percentages mentally at the bedside or in the clinic or surgery (or, indeed, in the laboratory).

We can use SD to calculate RCV. The laboratory's precision is derived from the internal QC program of the laboratory at the clinical decision-making level. This could be the mid-point of the reference interval for analytes such as sodium, chloride, potassium, or calcium when low and high results are equally interesting clinically; the upper reference limit when high results are most interesting, for example, for enzyme activities such as Alk Phos, ALT and amylase, and bilirubin; or the lower reference limit when decreases in analytes are of most clinical concern, for example, in albumin. When fixed limits are used for interpretation, such as for cholesterol, these should be used as the relevant clinical decision-making level.

It is assumed that the CV_A used in the formula is a constant, which, of course, it might not be in reality, particularly if there were usually significant changes in bias on re-calibration. We do not want to keep re-evaluating and changing our RCV. Therefore, this is another reason for minimizing precision and for trying hard to eliminate changes in bias over time. Then, the CV_I gleaned from the database could be easily converted into SD at the level chosen as the decision-making point as SD = [CV * level]/100. Then, generate the RCV in units.

At one time, we used this approach, calculated the RCV in unit terms, and reproduced a comprehensive listing of change in units at various levels of probability, in our own Laboratory Handbook, a part of which is shown in Table 3.8.

I suspect that this is too complex for most users and that a simple listing of changes at 95% and 99% probability—calling these, for clinical purposes, signifi-

Table 3.8 Significance of Changes in Serial Results: Example of a Comprehensive Table for User Guide or Laboratory Handbook

Analyte	Units	Probability that Rise or Fall Is Significant (%)					
		60	70	80	90	95	99
Albumin	g/L	<1	1	2	3	4	6
Alk Phos	U/L	4	9	13	20	26	36
ALT	U/L	3	5	8	13	16	23
Amylase	U/L	2	5	8	13	16	23
Bilirubin	µmol/L	<1	1	2	3	4	6
Calcium	mmol/L	0.02	0.04	0.06	0.09	0.12	0.17
Chloride	mmol/L	1	2	3	4	5	7
Cholesterol	mmol/L	0.2	0.3	0.5	0.8	1.1	1.5

cant and highly significant—would have many advantages and would be used more frequently.

Graphic Approaches. Often a graphic aid gives the same information as a table of numbers but in a more user-friendly format. It is easy to generate a graph of probability of change against change for any particular analyte (the probability for any particular change can be calculated easily by rearranging the RCV formula). A graph of probability (that change is significant against percentage of change for cholesterol) is shown in Figure 3.4. We assumed that CV_A was 3.0% and that CV_I was 6.0%.

Such a graph could be calibrated in units of mmol/L, using SD as shown earlier. We might make this most useful by doing all the calculations at the fixed clinical decision-making limit of 5.0 mmol/L. The problem with this approach, however, is that the change in units calculated for the y-axis would only be really accurate for changes from 5.0 mmol/L used on the x-axis. For graphs, using percentages and CV has more advantages.

The merit of these graphic aids is that they are easy to use. In addition, they demonstrate the very important concept that the probability that a change is significant is not directly proportional to the change. Doubling the change in results does not make the change twice as likely to have occurred.

The main disadvantage is that we can really only expect that a few of these would be used by our clinicians for analytes of particular interest to them.

RCV in Interpretation of Reports It is easy to prepare tables for laboratory handbooks or user guides, and easy to prepare graphs of probability that change is significant versus percentage change. However, these have disadvantages. We

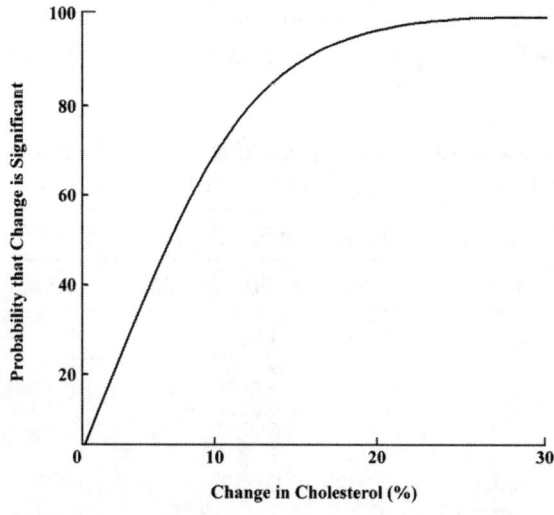

Figure 3.4 Graphic Relating Change in Results to Probability

CHANGES IN SERIAL RESULTS

need to use these RCV in everyday practice just as we frequently use conventional population-based reference values, and as we use fixed limits when it is clinically appropriate to do so.

In the laboratory in Ninewells Hospital and Medical School, Dundee, Scotland, we have used "flags" on our reports of laboratory data to help clinicians interpret laboratory results (see Figure 3.5).

We use the following "flags":

- \> means higher than the upper reference limit numerically, but this may not be of clinical importance,
- < means lower than the lower reference limit numerically, but this may not be of clinical importance,
- \>\> means higher than the upper reference limit numerically and likely of clinical importance,
- << means lower than the lower reference limit numerically and likely of clinical importance,

BIOCHEMICAL MEDICINE

DMR 374

NINEWELLS HOSPITAL AND MEDICAL SCHOOL   Telephone 660111 Ext 32601/2

Name: Sex: PID   DoB:
                                                         Lab No:
N/W Ward 20   Clinician: Dr

Analyte	Value	Flag	Units	Reference
SODIUM	138	*	mmol/L	(135-147)
POTASSIUM	5.0		mmol/L	(3.5-5.0)
UREA	9.5	**	mmol/L	(3.3-6.6)
CREATININE	137	>	umol/L	(50-100)
ALT	19		U/L	(12-41)
BILIRUBINS	100	>>	umol/L	(0-15)
ALKALINE PHOSPHATASE	130	>	U/L	(20-80)
ALBUMIN	23	<<	g/L	(36-50)
CALCIUM	2.27	**	mmol/L	(2.10-2.55)
CALCIUM (CORRECTED)	2.68	*	mmol/L	(2.10-2.55)
MAGNESIUM	0.82	**	mmoL/L	(0.70-1.15)
PHOSPHATE	0.93		mmol/L	(0.90-1.70)
C-REACTIVE PROTEIN	192.0	*	mg/L	(up to 5)

Lab.Comments: Sample Date/Time
                                                         30 Aug 2000

Request Entered: 30 Aug 2000 08:51   Report Printed:
                                                         REPORT RECEIVED
                                                         DOCTOR'S INITIALS

Figure 3.5 Example of a Laboratory Report Showing Use of RCV

These results are from a 50-year-old woman in our intensive care unit who had suffered a severe subarachnoid hemorrhage. As recommended, we report the analyte name in full, the result, the units used, and the age and sex matched population-based reference intervals.

- ∗ means significant change (95%), and
- ∗∗ means highly significant change (99%).

These flags are inserted by the LIMS. The ∗ and ∗∗ flags are calculated from the RCV formula as described earlier.

Using flags of this type has some significant advantages.

- It helps users select important information from the vast amount generated on most patients, and
- it educates clinicians about change in serial results, showing that
  - significant changes can occur when serial results are all within the reference interval,
  - a change in numerical results is not necessarily significant, and
  - results can change from inside to outside reference intervals (and vice versa) without significance.

Using flags of this type has a philosophical disadvantage: the flags usually adopted by laboratories and available in most LIMS, such as high, low, and usual (not flagged) actually simply divides the numerical data we have worked hard to produce with good precision and low bias into three classes. Our Dundee system divides the data into more classes than this—but it could still be said that we are degrading good numerical data by encouraging simplistic interpretation.

On balance, we think that objectively informing users of the significance of changes in numerical laboratory results is of clinical value.

Delta-Checks and RCV Patient data can be used in a number of ways to monitor the quality of the various processes involved in generating a test result. These are very well described by Cembrowski and Carey and include "delta-checks." Most LIMS now allow implementation of some type of delta-checking (Δ-checking).

The fundamental basis of Δ-checking is as follows.

- If patients are stable, the changes in results—the delta-values (Δ-values)—should be small.
- If the Δ is large, and greater than a pre-defined limit, then there is a Δ-check failure, and
  - the Δ-check failure may be due to real change in the patient or errors associated with samples (either the first or the second).

Delta-check values are traditionally generated in two ways. The first is "scientific" and involves collecting large numbers of consecutive pairs of patient data that are like the patient data to which we will apply the Δ-check values. Then, we calculate and plot the Δ-values in a frequency distribution histogram; we also calculate the Δ-check values that will be used to highlight either 5% or 1% of Δ-values. The second approach is to set Δ-check values based empirically on experi-

CHANGES IN SERIAL RESULTS

ence and then adjust them with time so as not to generate too many Δ-check failures.

As described earlier, there are considerable advantages to using RCV to alert users that serial results in an individual have changed significantly (95%) or highly significantly (99%). We can also use RCV in Δ-checking: we use exactly the RCV clinical flags as the Δ-checks in our laboratory operation in Dundee. We can calculate the RCV as percentage of change using CV and as numerical change using SD. Most LIMS can accept Δ-check values in either of these formats.

We use our LIMS to flag the results as * for significant change and as ** for highly significant change. Results flagged with ** (and with << or >>) are held for review by a clinical biochemist. Results flagged with * (and results with < or > or not flagged results—those within the reference interval and unchanged significantly) are reported to the user without intervention. This automatic reporting of usual, slightly unusual, and 95% changed results, called "exemption reporting," has reduced the number of results that require investigation.

When a result is flagged as **, then it has failed the Δ-check. The patient may have changed or there may be an error in either result. Cembrowski and Carey have generated an excellent comprehensive algorithm for investigating Δ-check failures. A very simple version showing the basic steps is presented as Figure 3.6.

Essentially, if the Δ-value is compatible with the clinical situation, then we report the most recent result. If the Δ-value is incompatible with the clinical situation, then the sample—preferably from the original sample tube taken from the patient rather than an aliquot—should be re-analyzed. If the unusual Δ-value is confirmed, the change will be highly significant according to the RCV, and dis-

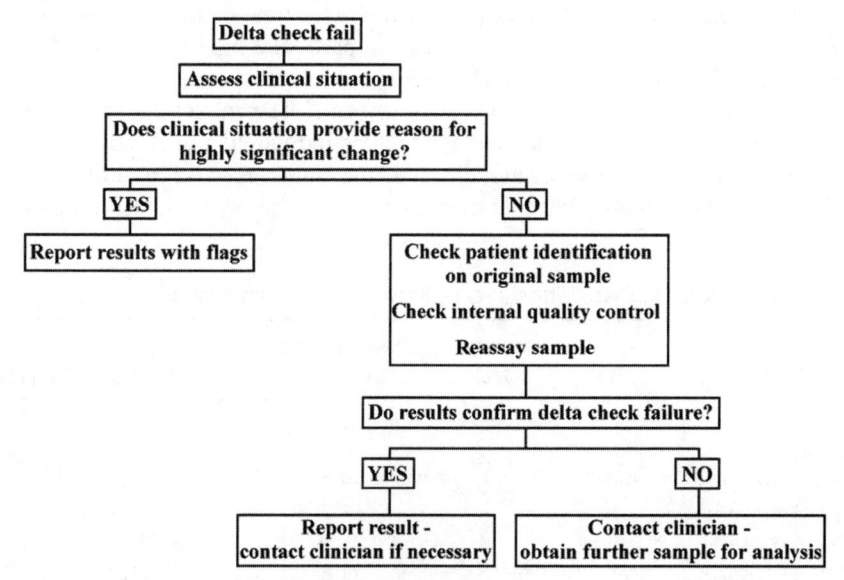

Figure 3.6 Simple Algorithm for Delta Check Failure

cussion between the laboratorian and test requestor is indicated. Taking a new sample should often be recommended to resolve the apparent conflict between clinical situation and laboratory test result.

It is highly recommended that we should harness the current capacity of most available LIMS to deal with Δ-checks. We should use RCV at 99% probability as our Δ-checks, and get our LIMS suppliers to adapt their Δ-check software to allow changes in serial results to be flagged on the laboratory data reports rather than simply making them available to laboratory staff for quality management purposes.

Time Series Analysis Earlier, we looked at changes in liver function test in a 63-year-old man who was prescribed statin therapy to lower his less-than-ideal serum cholesterol concentration.

We saw that the test results over time were not constant. We expected random variation. It looked as if the serum AST activity was rising. We calculated the RCV for serum AST activity as 33.2% for 95% probability—significant change—and 48.3% for 99% probability—highly significant change.

Results of the serum AST activity assays are shown in Table 3.9. The significance of changes between successive pairs of serial results is also shown. Even although the AST activity results rise time after time from a "baseline" of 16 U/L to 69 U/L after 8 months, two of the changes are not significant, two are significant, and only one is highly significant.

But, there is clearly a trend upwards and, if we compared values at 4, 6, and 8 months to the baseline, the changes would be highly significant. The question then arises: "Can we adopt more sophisticated techniques than RCV to assess changes with time?".

Mathematical methods for reviewing an individual's serial results have been developed. These are sometimes called methods for time series analysis. Eugene Harris has looked at these methods and two in particular may be worthy of attention, since these can be applied to short series of results.

One model is called the "homeostatic model." The model assumes that the analyte behaves in a random manner, and that the individual has a homeostatic setting point. In contrast, the other model, called the "random walk," assumes that the analyte behaves randomly but that there is not a true homeostatic setting point.

Table 3.9 Serum AST Activities in a Patient Treated with Statin

Time	AST (U/L)	Change (U/L)	Percentage change	Significance	Flag
Pre-dose 1	16	—	—	—	—
Pre-dose 2	18	+2	12.5	< 95%	none
After 2 months	24	+6	33.3	> 95%	*
After 4 months	40	+16	66.7	> 99%	**
After 6 months	54	+14	35.0	> 95%	*
After 8 months	69	+15	27.8	< 95%	none

CHANGES IN SERIAL RESULTS

These models have not been widely applied. Those who have attempted to use them on apparently healthy people, for example those undergoing regular health assessment, have found that there are very many "false positive" results. Many results are flagged as having changed significantly but there is no indication that state of health has changed and no evidence of the presence of disease. The general consensus seems to be that these methods are too complex and too time consuming to be used in everyday laboratory or clinical practice.

SUMMARY

We have covered the following key points in this chapter.

- Test results are used for many purposes, most for monitoring individuals over time.
- Monitoring involves assessing differences in an individual's serial results. The use of either conventional population-based reference values or fixed limits has disadvantages.
- The total variation of any result is due to pre-analytical, analytical, and biological sources. The total variation is found by adding variances—the squares of SD—or the squares of CV if the mean is constant for the components.
- Pre-analytical variation should be minimized by adopting strict protocols for patient preparation, and sample collection, transport, and handling. Education of staff, institution of good laboratory practice, and adherence to standard operating procedures are vital.
- Reference change values (RCV) are simple to calculate and are the most appropriate tools to assess changes in serial results. It is common to use 95% probability for significant change and 99% probability for highly significant change; the appropriate values for Z are 1.96 and 2.58 respectively.
- The probability that any particular change has occurred can be calculated by rearranging the RCV formula.
- Many terms are used in medicine to describe clinical decision-making. Clinically important probability may not always be at 95% or 99%. Lower probability for change may be very important clinically.
- RCV can be lowered by improving precision. The probability that a change is significant will be higher with better precision. RCV and probability are not linearly related to precision.
- RCV will vary with changes in analytical precision and bias. These analytical variations should be eliminated if possible and certainly should be minimized by careful attention to quality management practices, particularly at re-calibration.
- RCV have some theoretical disadvantages: most data on CV_I have been found on healthy people, so using conventional RCV on ill people will lead to "false positives." If tests are performed frequently, serial results will not be random and will be auto-correlated, leading to "false negatives." The advantages of using RCV as aids in clinical interpretation outweigh the disadvantages.

- Laboratories can provide RCV in User Guides or Laboratory Handbooks. For individual analytes, graphs of probability of change against change are easy to produce, although it is unlikely that these will be widely used.
- Laboratories should flag significant and highly significant changes on their laboratory reports.
- Delta-checking is widely available. Delta-check values are most often derived empirically. Objectively calculated RCV could be used for delta-checking laboratory data as well as for assisting interpretation.
- It is highly recommended that the current capacity of most available LIMS to deal with delta-checks should be harnessed to allow the data to be used for reporting results with RCV applied to flag results.
- Models are available for time series analysis but these have not been widely applied: they give many false positives and are too complex and too time consuming for everyday practice.

FURTHER READING

1. Cembrowski GS, Carey RN. Laboratory quality management: QC=QA. Chicago: ASCP Press, 1989.
2. Fraser CG, Harris EK. Generation and application of data on biological variation in clinical chemistry. Crit Rev Clin Lab Sci 1989;27:409–437.
3. Harris EK, Boyd JC. Statistical bases of reference values in laboratory medicine. New York: Marcel Dekker Inc., 1995.
4. Harris EK. Some theory of reference values. II. Comparison of some statistical models of intra-individual variation in blood constituents. Clin Chem 1976;22:1343–1350.
5. Harris EK. Statistical aspects of reference values in clinical pathology. Prog Clin Pathol 1981;8:45–66.
6. Lacher DA, Conelly DP. Rate and delta checks compared for selected chemistry tests. Clin Chem 1988;34:1966–1970.
7. Queralto JM, Boyd JC, Harris EK. On the calculation of reference change values, with examples from a long-term study. Clin Chem 1993;39:1398–1403.
8. Ricos C, Alvarez V, Cava F, Garcia-Lario JV, Hernandez A, Jimenez CV, Minchinela J, Perich C, Simon M. Current databases on biologic variation: pros, cons, and progress. Scand J Clin Lab Invest 1999;59:491–500. Available online at www.westgard.com/guest17.htm.

Chapter 4
The Utility of Population-Based Reference Values

Laboratory test results are used for many purposes, and therefore we must be able to objectively interpret the numbers we report from our laboratories, irrespective of the clinical use made of the test results. This interpretation often involves comparing test results against

- population-based reference values,
- locally agreed-upon protocols for clinical action,
- values proposed by expert individuals, groups, or committees,
- values based on outcomes, for example, risk,
- multiples of the upper reference limit, or
- previous results obtained on the individual.

Population-based reference values remain the most commonly used interpretative aids in diagnosis and case finding. Many users of test results apply them for other purposes, as well, without realizing their many deficiencies. In this Chapter, we examine the development and true clinical utility of population-based reference values.

Population-based reference values have enjoyed renewed professional interest recently, along with quality specifications in laboratory medicine. Some of the problems mentioned in a recently published opinion piece on the need to revisit traditional approaches will be addressed in this chapter. Moreover, the National Committee for Clinical Laboratory Standards (U.S.) recently published a very comprehensive document on defining and determining reference values in the clinical laboratory.

THE "REFERENCE VALUE" CONCEPT

The term "reference values" is preferred to the more common term "normal range." The term "normal" has many uses in medicine. In a statistical context, normal values means that the values are distributed in a bell-shaped symmetrical Gaussian distribution. Many analytes of interest in laboratory medicine, however, do not have this type of distribution. Some have recommended that the term Gaussian distribution be used in preference to the term normal distribution.

In a clinical connotation, normal values could mean that the values indicated the individual was healthy and the risk of having disease was low. Sometimes, such as for cholesterol—certainly in Scotland—the values found in the population are much higher than those associated with low risk of disease. In these cases, the term normal might be replaced by usual.

In its epidemiological meaning, normal values could mean that the values found were typical of the values found in the population at large. Normal in this situation means that ranges are created and results outside these ranges are called abnormal. Unusual or atypical are better terms than abnormal in these situations, because these do not signify that such values may be harmful.

The IFCC discourages using the terms normal range, normal values, and normal reference values, has worked on replacing the problematic term normal values with the more appropriate term reference values, and has published a series of six approved recommendations, covering

- the concept of reference values,
- selecting individuals for the production of reference values,
- preparing individuals and collecting specimens for producing reference values,
- controlling analytical variation in the production, transfer, and application of reference values,
- statistical treatment of collected reference values—determination of reference limits, and
- presenting observed values (test results) related to reference values.

The International Committee for Standardization in Haematology, the World Health Organization, and NCCLS have all approved these six far-reaching recommendations. In this Chapter, we explore these recommendations, and the ramifications of biological variation on the derivation and use of reference values.

REFERENCE VALUE TERMINOLOGY

A *reference individual* is an individual selected for comparison using defined criteria. The IFCC does not say that reference individuals have to be healthy.

A *reference population* consists of all possible reference individuals. This is a rather theoretical group since the reference population will usually consist of an unknown number of people.

A *reference sample group* is an adequate number of reference individuals taken to represent the reference population. This group represents the reference population and so, at least in theory, would be chosen at random from this population. In practice, we hope that the selected reference individuals who make up a reference sample group represent the reference population as a whole.

Reference values are the values obtained on reference individuals for an analyte. Reference values will not be identical for all reference individuals; they will have a statistical dispersion termed the *reference distribution*.

We then calculate *reference limits*, created so that a pre-set fraction of the reference values are less than or equal to the limits.

THE UTILITY OF POPULATION-BASED REFERENCE VALUES

    reference individuals
        make up a
        reference population
            from whom are selected a
            reference sample group
                on whom are determined
                reference values
                      on which is observed a
                      reference distribution
                            on which is determined
                            reference limits
                                that define a
                                reference interval

Figure 4.1   Reference Value Concept

The upper and lower reference limits are used to define the *reference interval*.

The relationships between the various terms are shown in a flow chart of words format in Figure 4.1.

The IFCC recommends the term *reference interval* rather than *reference range*. Pedantically, a range is actually the difference between two numbers, so if the reference limits for serum sodium are 135 and 147 mmol/L, the reference range is 12 mmol/L and not 135–147 mmol/L.

SELECTING REFERENCE INDIVIDUALS

The IFCC only allows us to determine reference values directly on reference individuals. A considerable body of literature provides strategies for indirect selection of reference individuals—i.e., using mathematical or graphical ways of looking at routine laboratory results and deciding which part of the set of available numbers belongs to the "normal" component of the population. An elegant recent work from Stockholm, Sweden, suggests that samples from general practitioners might be very useful in this regard.

Some recent publications show how reference values can be obtained through co-operation among laboratories using the same methodology. Those who favor this approach suggest that it is cheaper and easier than using the IFCC-recommended approaches, which do, indeed, take considerable time and trouble.

Using the permitted "direct" approach, then, there are two ways to select reference individuals who make up our reference sample group who, we hope, represent the reference population. The IFCC calls them the *a priori* (before) and the *a posteriori* (after) approaches. The *a posteriori* approach means that we take a large amount of data from those who have been medically examined, i.e., from people who attend for health monitoring or preventive medicine or "well person" examinations. We then apply pre-set exclusion or inclusion criteria to make sure that only reference individuals are included in the data-handling steps.

Few researchers have access to the large number of people required for this technique, but some have published very useful sets of reference intervals derived in this manner, i.e., values derived from 31,000 healthy women and 51,000 healthy men examined at BUPA Medical Research in the U.K. for common biochemical and hematological analytes.

The *a priori* approach is favored, whereby we select reference individuals from the population and apply inclusion or exclusion criteria. We could use apparently healthy people, either formally assessed, superficially investigated, or not examined in any way for their state of health. We could also use diseased people provided the disease they had did not affect the studied analytes, but this is less commonly done.

Most often, we have to be pragmatic and use relatively young, ambulatory and available people such as students, hospital staff, or factory workers as our reference subjects. The NCCLS document suggests that we should use well-designed questionnaires when selecting reference individuals. Questionnaires should be simple and not intimidating, and provide information not only for deciding whether to exclude or include the individual from the reference sample group but also to determine data such as age and sex that might be useful in later stratification or partitioning. The NCCLS guidelines include a sample questionnaire that might be used as a basis for this approach.

Regardless of which reference individuals we select, we should apply the exclusion criteria exactly. Individuals should

- be ostensibly healthy,
- not be taking oral contraceptives,
- not be taking prescribed or over-the-counter drugs,
- not consume excessive amounts of alcohol, and
- preferably not use tobacco products or other such recreational substances.

How many reference individuals do we need to develop satisfactory reference values? The IFCC suggests 120. When we estimate the reference limits statistically, the numbers we calculate do have inherent variability. An estimate of this variability can be calculated as the confidence interval. Theoretically at least, we should know the confidence intervals of the reference limits. The NCCLS guideline provides details about calculating this figure using a very simple non-parametric technique, yet confidence intervals are rarely documented on reports of test results.

The confidence intervals of the reference limits depend on the number of reference values, and the recommended "120" was selected to make these suitably small. This number of reference individuals may be difficult to obtain, particularly if age and/or sex stratified values are needed. However, there is nothing mandatory about this particular number: reference values can be generated from a larger

THE UTILITY OF POPULATION-BASED REFERENCE VALUES

or smaller reference sample group. Only the confidence intervals of the reference limits will be smaller or larger respectively.

Sample Collection, Handling, and Analysis We should standardize the sample collection and handling techniques adopted to minimize the many pre-analytical sources of error. Standardization includes the following:

- taking samples at the same time of day, usually in the morning, unless we are developing reference values specific to other times,
- ensuring that the reference individuals have samples taken under the same conditions: a limit of one unit of alcohol consumed the day before, the usual previous food intake, no strenuous exercise before sampling, no food and only one glass of water during the previous 10 hours, and sitting down for 15 minutes prior to sample collection,
- taking blood samples with a standard phlebotomy technique, preferably by experienced phlebotomists, into the same type of collection tubes, preferably without stasis,
- transporting the samples to the laboratory under the same conditions of temperature and elapsed time, and
- centrifuging, when required, at the same speed for the same period of time and at the same temperature.

Sample collection and handling principles should be consistent whether we are studying blood or whether we are studying fluids other than blood for reference values. For example, if we were studying analytes in 24-hour urine samples, then the instructions given to the subjects should be identical, the start times and stop times should be standardized, and the amount of stabilizer or preservative should be constant.

Since reference values are applied over many analytical runs in everyday practice, we do not need to store samples as was advocated for studies on biological variation. We should analyze samples in a number of routine runs, applying the same principles to the sample analysis. We should analyze the samples with the method running under quality controlled but realistic conditions. We should try to ensure that the method does not have any bias. (This should, of course, apply to all methods used in laboratory medicine.) We should keep the precision at the same standard as analysis done under good control with good laboratory practice and good quality assurance techniques in force.

INVESTIGATING OUTLIERS

Once we have generated the test results from our reference sample group, we must examine the data to determine the clinically most useful reference limits and reference interval. Before we do this, however, we must examine the data for outlying values that were clearly not from the reference distribution. If such values are included, we would generate artefactually wide reference intervals, clearly limiting their clinical utility.

The first procedure is to draw a graph of frequency of results (on the y-axis) against test results on the x-axis). We must look at data before entering the whole set of numbers into a statistical computer program. Visual inspection of the graph will let us see quite quickly what the distribution looks like and whether there are outliers.

Those who generate reference values often have real trouble deciding if any of the results are outliers, but it can be done very simply. We use Reed's criterion as we used in Chapter 1 when considering generation of the components of biological variation. This very simple test considers the difference between the extreme value and the next lowest (or highest) value, and rejects the extreme value if this difference is more than one-third of the absolute range of values (highest to lowest).

Sometimes there are two or three values that are rather different (low or high) compared to the main bulk of the reference values. In these situations, the traditional sequential application of Reed's test may not pick out these values as outliers. Then we must test the value that looks least like an outlier first. If this is an outlier (difference between it and the next value > one-third of the range between it and the extreme value at the other end of the distribution), then all potential outlier values are rejected.

DERIVING REFERENCE LIMITS AND INTERVALS

The IFCC recommendations describe the statistical methods in great detail. Of the two approaches—the parametric and the non-parametric—the IFCC recommends the latter method, which is simple and makes no assumptions about the type of distribution. For example, if we wish to calculate a reference interval that includes 95% of the population, which is the most often used (but only by convention), we would perform the following steps.

- Put the values in ascending order and give each value a consecutive number: 1, 2, 3, and so on to the last. Call this last number n.
- Calculate the number of the value that is equivalent to the lower reference limit—$0.025 * (n + 1)$.
- Find the value equivalent to this number. If the number is a whole number this is easy. If it is a number plus a fraction, then interpolate between the two appropriate values. This is the lower reference limit.
- Calculate the number of the value equivalent to the upper reference limit—$0.975 * (n + 1)$.
- Find the value equivalent to this number. If the number is a whole number this is easy; if it is a number plus a fraction, then interpolate between the two appropriate values. This is the upper reference limit.

The parametric approach, which requires examining the data to see if the distribution is Gaussian, is far more complex and not really recommended. It can be

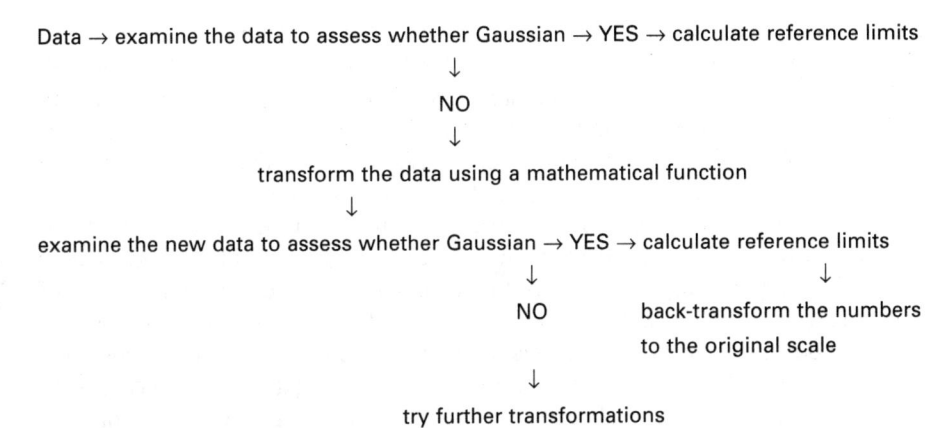

THE UTILITY OF POPULATION-BASED REFERENCE VALUES

Data → examine the data to assess whether Gaussian → YES → calculate reference limits
↓
NO
↓
transform the data using a mathematical function
↓
examine the new data to assess whether Gaussian → YES → calculate reference limits
↓ ↓
NO   back-transform the numbers
     to the original scale
↓
try further transformations
↓
examine the data to assess whether Gaussian → YES → calculate reference limits
↓
back-transform the numbers
to the original scale

Figure 4.2 Parametric Approaches to Data Handling for Calculating Reference Limits

done, however, using a variety of statistical tests including "goodness of fit" tests such as the Anderson-Darling or Kolmogorov-Smirnov tests, or "coefficient" tests that examine the shape of the distribution by calculating skewness and kurtosis. The IFCC recommendations, and the excellent and readable review of Solberg and Grassbeck, describe these tests in detail. If the distribution is not Gaussian, then we might be able to transform it into this type of distribution. The most common transformation uses logarithms, but many others are possible. The parametric procedure is shown in Figure 4.2, but the non-parametric approach remains much easier.

FACTORS AFFECTING REFERENCE VALUES

These factors can be classified into five groups: endogenous, exogenous, genetic/ethnic, laboratory, and statistical.

Endogenous factors are inherent in individuals and cannot be modified. The two most important endogenous factors are age and sex. Many analytes do change over the lifespan (see Chapter 1). Chronological age and biological age, however, may not be the same. In addition, reference values for many analytes depend on sex, particularly during the female's reproductive phase.

Exogenous factors include fasting; starving; exercise; living at high altitude; pregnancy; habitual use of alcohol or other so-called recreational substances; body size—can affect certain test results such as serum creatinine; and medical or surgical care, including immobilization and drug administration.

Genetic or ethnic factors might be important depending upon the location of the laboratory and the population served. Many documented influences in different groups, however, may actually be due to exogenous factors. For example, serum immunoglobulin concentrations are higher in peoples living in Africa, but fall if they move to European countries. Immigrants' serum cholesterol results change to be more like the native population; some have suggested that this might be due not only to diet and lifestyle, but also to the amount of winter sunshine.

Laboratory factors that can affect results include the type of sample used (serum or plasma); the standard phlebotomy technique used (with/without tourniquet); the time taken for transport to the laboratory; the time and speed of centrifugation; and the actual methodology used. One example in which laboratory methodology affects reference values is the measurement of lactate dehydrogenase activity. Methods that measure the pyruvate-to-lactate reaction have activities much higher than methods that measure the lactate-to-pyruvate reaction.

The *statistical approach* we take to determine reference limits can also affect the reference intervals used in the interpretation of test results.

STRATIFYING (OR PARTITIONING) REFERENCE VALUES

Although specialist texts often give the impression that stratification is a complicated matter, it is actually rather straightforward and may prove to be of significant clinical value.

Sinton and co-workers advocate the simplest technique. When is stratification necessary? Separate stratified reference intervals are not required unless the difference between the means of the potential sub-groups exceeds 25% (one quarter) of the 95% reference interval for the entire non-stratified group. For example, for serum creatinine, the reference interval for the entire reference sample group was 50–120 µmol/L. The mean for women was 75 µmol/L and that for men was 95 µmol/L. Then, because $95-75 = 20$ which is greater than 17.5 (calculated as 25 % of $120-50$), stratification according to sex is recommended.

For serum sodium, however, the reference interval for the entire reference sample group was 135–145 mmol/L. The mean for women was 140 mmol/L and that for men was 139 mmol/L. Then, because $140-139 = 1$ which is less than 2.5 (calculated as 25 % of $145-135$), stratification according to sex is not required.

A more complex technique that examines the width of distributions as well as the means has been advocated by Harris and Boyd, and recommended by NCCLS, among others. The method looks difficult, but is actually easy to calculate.

- First, calculate Z as

  Z = (mean of group 1 − mean of group 2)/
  $(SD_1^2/\text{number in group 1} + SD_2^2/\text{number in group 2})^{1/2}$.

  This is the simple standard normal deviate test.

- Then see whether the calculated Z exceeds the critical statistical value derived by Harris and Boyd.

THE UTILITY OF POPULATION-BASED REFERENCE VALUES

We can easily calculate the critical value of Z as the following number:

3 [(number in group 1 + number in group 2)/240]$^{1/2}$,

that is, 3 times the square root of the total number of reference values divided by 240. If Z is larger than the critical value, then stratification is warranted.

- Next, investigate the relative magnitude of the standard deviations of the potential sub-groups. If the larger SD is greater than 1.5 times the smaller, then stratification is warranted.

PROBLEMS WITH CONVENTIONAL REFERENCE VALUES

As noted earlier, current dogma is that *every laboratory should generate its own reference values*. Literature from manufacturers and suppliers of reagents and analytical systems almost ubiquitously state this recommendation. Laboratories should also supply reference intervals with every result, or at least list these in detail in the Users' Guide or Laboratory Handbook.

Even though the IFCC offers a variety of options on presenting reference values in Part 6 of its recommendations, most laboratories simply give the numerical reference intervals. As professionals interested in quality interpretation as well as analytical quality, we should take time to discourage users of test results from using textbooks, handbooks of "normal values," diaries, promotional material from diagnostic and drug suppliers, and other similar sources.

In addition, for important factors such as age and sex, we may need to stratify the data into sub-sets of reference values using the relatively simple tests detailed above. If we stratify the data we have generated into many small groups using the recommended techniques, we will have very small numbers of reference individuals in each group, which will lead to larger confidence intervals (more uncertainty).

The recommended minimally adequate number required to generate reference values, 120, must theoretically, at least, be multiplied by the number of stratified groups deemed necessary. As you can see, generating reference values takes considerable time and effort, and organizational, analytical, and statistical skills. Is this productive use of scarce laboratory resources?

Since most test results are used for monitoring patients, and not for diagnosis, population-based reference values are thus generally inappropriate as monitoring tools. Our interpretation should be concerned with whether or not a change has occurred. We should use RCV based on analytical and within-subject biological variation, which are simple to calculate from easily obtainable data.

Most quantities have marked individuality, so individuals can have values that are highly unusual for them but still lie within reference limits. In consequence, reference values are not good criteria in case finding or in screening to detect early, latent, or sub-clinical disease.

Users generally understand that the reference interval comprises 95% of the population and, by definition, 2.5% of people are outside each of the reference limits. In addition, but not as appreciated, the more tests that are performed on a subject, the greater the chance of finding an unusual result.

Experienced clinicians do not use only reference values to make decisions; they apply judgments gained from experience. However, they do require guidance about the values found in health and disease.

Local guidance and protocols are very often based upon professional experience and clinical judgment rather than strictly on reference values.

Decision-making, even using the clinical laboratory alone, is not usually based on a single result but is in reality a mental multivariate analysis—and laboratories do not provide multivariate analyses of data.

Interpretation is often based on fixed criteria laid down by expert national or international professional groups (and developed from evidence-based medical principles) rather than on population values. Examples include cholesterol and other lipids, glucose, therapeutic drugs, and substances taken in overdose such as acetaminophen (paracetamol).

However, even though population-based reference values do have disadvantages, we should have a strategy for developing them in our laboratories.

GENERATING REFERENCE VALUES—A PRAGMATIC APPROACH

The factors listed above suggest that it is not really necessary to develop reference values "from scratch" for the entire test repertoire we offer in our laboratories. We therefore propose a pragmatic approach to their development.

We can use the following hierarchical cascade of approaches to derive reference values, adopting the strategy that most closely approaches the ideal in terms of feasibility, practicability, and cost effectiveness.

1. If possible, use the strategy published by the IFCC Expert Panel on Theory of Reference Values as documented here and in published reviews and recommendations.
2. Use the above approach but with a less stringent selection of the reference sample population; for example, laboratory staff, hospital staff, students, blood donors, or patients without problems likely to affect the test under study.
3. Use specific literature giving good peer-reviewed data on reference values, particularly individual publications that use the particular methodology adopted in the laboratory.
4. Use books that contain well-researched information on reference values. These are particularly useful for reference values in the neonate, the child, and the elderly, because it is difficult for laboratories to develop these for themselves. In addition, use these sources to directly and proportionately extrapolate the data developed from the first three approaches if local age and sex stratified reference values are desired.

THE UTILITY OF POPULATION-BASED REFERENCE VALUES

5. Use other literature that gives information on reference values or other strategies for interpreting laboratory test results.
6. Then use manufacturer's data, particularly for the more esoteric tests.

For approaches 3–6, we should do objective studies on whether the reference values can be transferred to the laboratory.

TRANSFERABILITY OF REFERENCE VALUES

Because of the very real difficulties in generating good quality reference values, laboratories must often transfer data from other sources. Such sources include values created previously on a different analytical system, values from another laboratory, values from peer-reviewed literature, values from books, and even values from the literature provided by manufacturers. Regardless of the source, the data should be validated before being put into practice. Recent NCCLS guidelines provide the best set of recommendations on how to do this. There are three possible techniques.

1. We can subjectively assess the validity of the transfer by inspecting the detail of the original data. We can examine all factors that affect reference values—endogenous, exogenous, ethnic/genetic, laboratory, and statistical. If the factors are consistent between the laboratory and the source, then the laboratory can accept these without further experimental work. The laboratory should, of course, carefully document the sources of all its reference values.
2. This second technique appeals to the clinical scientist because it is simple, easy, and less subjective than technique #1. The laboratory carefully chooses a small number of reference individuals (20) and generates reference values. Then the laboratory examines the data for outliers using Reed's criterion. The laboratory discards any outliers, and replaces them, to keep the number at 20. Then, if no more than 2 of the 20 results lie outside the reference interval, the laboratory can accept that interval. If 3 or more fall outside, then the procedure should be repeated with another 20 reference individuals. On this second pass, if no more than 2 of the results lie outside the reference interval, the laboratory can accept the reference interval. But, if more than 3 are outside, a different source of reference values is clearly required.
3. This approach involves conducting a small study according to IFCC recommendations. Using 60 reference individuals that are truly representative of the reference sample population, we then compare the reference values to the set we wish to transfer. Using the methods of Harris and Boyd, we compare both the means and standard deviations of the two reference value sets.

A fourth approach uses data from method comparison studies regardless of whether they were done as part of a simple validation or as an extensive evaluation. This technique (advocated by many, and discussed but not approved in the excellent NCCLS guideline), is good if the laboratory had a well-established method, had created good reference values, and wished to transfer these data when commissioning a new instrument or introducing a new methodology. The method comparison studies would have given a simple mathematical relationship of the following type:

new results = a factor times old results + intercept,

that is, a linear regression equation

$$y = a * x + b,$$

where a is the proportional bias, which may be more or less than 1.0, and b is the constant bias, which may be positive or negative. We can easily calculate reference values for the new method from the existing reference values and this simple equation.

Regardless of how we generate reference values, ongoing monitoring of the number of values that fall outside the reference limits may indicate the appropriateness of the values. Further, any comments or complaints from users regarding the number of unusual results or the number of clinical impressions not borne out through laboratory investigation should prompt a serious and detailed review of the reference values quoted by the laboratory.

THE INFLUENCE OF INDIVIDUALITY ON REFERENCE VALUES

Even though traditional reference values have many problems, many clinicians use these values to aid investigation and management as if they were "gospel."

A major problem is due to individuality. Serum creatinine provides a good example. When we examined the variability of serum creatinine over time in a few men and a few women (see Chapter 1), we found that individuals really are individuals. This limits the utility of conventional reference values.

SERUM CREATININE IN AN ELDERLY POPULATION

Some years ago, we studied a cohort of 27 elderly people for a number of analytes, including creatinine. Figure 4.3 shows the means and absolute ranges we found for the 27 people. Subjects 1–13 were women and 14–27 were men.

The values for these women and men aged 70 and above, who were apparently healthy, lived independently, had no previous medical history of note, and fulfilled nearly all of the criteria for a reference sample group, appear higher than for younger people. Thus, age stratified reference values will most likely be required. Generally, women have lower values than men, but there is some overlap, because creatinine in serum in healthy people is mostly influenced by muscle bulk and, while women tend to be smaller than men, there are some large women and some small men. However, sex stratified reference values will probably be required.

THE UTILITY OF POPULATION-BASED REFERENCE VALUES

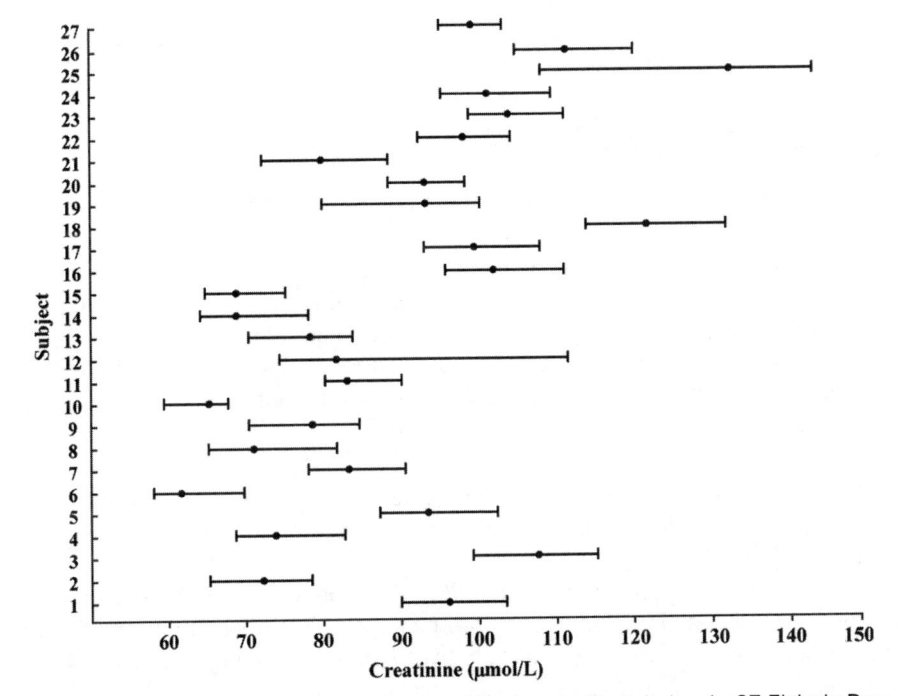

Figure 4.3 Means and Absolute Ranges for Serum Creatinine in 27 Elderly People (1–13: women; 14–27: men). (*Used with permission. From Fraser CG. Biological variation in the elderly: implications for the use of reference values. In: Faulkner WR, Meites S, eds. Geriatric clinical chemistry reference values. Washington, DC: AACC Press, 1994;44. Figure 4-1.*)

We had generated reference values stratified for age and sex and, at that time, quoted the following reference intervals for creatinine in individuals over 55 years: women: 60–98 μmol/L, and men: 66–128 μmol/L.

Further visual inspection of the figure demonstrates the following.

- No individual has values that span the entire reference interval.
- The range of values from each individual occupies only a small part of the dispersion of the reference interval.
- Most individuals have all values within the reference interval.
- The mean values of most individuals lie within the reference interval and are different from each other.
- A few individuals have values that span the lower reference limit; these individual have values that change from usual to unusual over time.
- A few individuals have values that span the upper reference limit; these individuals also have values that change from usual to unusual over time.
- One of the women has values for her mean and the range that fall outside the appropriate age and sex stratified reference interval.

- One of the men has a value for his mean that falls outside the appropriate age and sex stratified reference interval.

BIOLOGICAL INDIVIDUALITY

From the data in Figure 4.3, we drew these conclusions.

- No individual has values that span the entire reference interval.
- The range of values from each individual occupies only a small range compared to the rather large dispersions of the quoted reference intervals.
- Even if an individual's values spanned a reference limit or were all outside a reference limit, the range of values found for that individual was small compared to the reference interval.

Therefore, we know that the within-subject variation is relatively small, the mean values of the individuals differ from each other, and the between-subject variation is relatively large.

For creatinine, within-subject biological variation is less than between-subject biological variation. The estimate of within-subject biological variation (CV_I) for this particular group was 4.3% and the between-subject biological variation (CV_G) was 18.3%.

This can be expressed as $CV_I \ll CV_G$ and, when this holds for an analyte, the analyte is said to have marked *individuality*. We decide an analyte's individuality by calculating the *index of individuality*, usually abbreviated to II. This idea came from Eugene Harris, who suggested that we could calculate the index of individuality as the ratio of the total within-subject variation to between-subject biological variation—formally as

$$II = [CV_A^2 + CV_I^2]^{1/2}/CV_G.$$

This is most often simplified to

$$CV_I/CV_G$$

which is satisfactory if $CV_A < CV_I$, which it often is with current methodology and technology.

A *low index of individuality* means that the analyte has *marked individuality* whereas a *high index of individuality* means that the analyte has *little individuality*.

Some indices of individuality for commonly analyzed biochemical analytes are shown in Table 4.1.

Individuality is usual in laboratory medicine. This has important consequences for the use of conventional reference values, the use of laboratory tests, and the interpretation of laboratory test results.

THE UTILITY OF POPULATION-BASED REFERENCE VALUES

Table 4.1 Some Examples of the Index of Individuality (II)

Analyte	Within-subject variation (%)	Between-subject variation (%)	II
ALT	24.3	41.6	0.58
Albumin	3.1	4.2	0.74
Alk Phos	32.6	39.0	0.84
Bilirubin	25.6	30.5	0.84
Calcium	1.9	2.8	0.68
Chloride	1.2	1.5	0.80
CK	22.8	40.0	0.57
Creatinine	4.3	12.9	0.33
LD	6.6	14.7	0.45
Magnesium	3.6	6.4	0.56
Phosphate	8.5	9.4	0.90
Potassium	13.6	13.4	1.02
Protein	2.7	4.0	0.68
Sodium	0.7	1.0	0.70
Urate	8.6	17.2	0.50
Urea	12.3	18.3	0.67

BIOLOGICAL INDIVIDUALITY AND REFERENCE VALUES

The individuality of analytes significantly influences reference values. We have taken six of the men in whom we have studied both serum creatinine and serum iron. Figure 4.4 shows the mean and absolute ranges of these analytes in the six men.

The reference interval for elderly men for serum creatinine is 66–128 µmol/L. The reference interval for elderly men for serum iron is 5–32 µmol/L.

For creatinine, no one individual's values span the entire reference interval, and the range of values from each individual occupies only a small part of the dispersion. For iron, however, all individuals' values span almost the entire reference interval.

What happens if individuals become diseased and their results become unusual? For creatinine, individuals could have values that were very unusual for them but these clinically important results would still lie within the conventional reference interval. Generally, laboratories would not flag these results as unusual, because they use population-based reference intervals and not RCV. Moreover, many clinicians use reference values as important aids to interpretation, and they would not consider such results as significant in any way.

In contrast, for iron, if individuals had values that were even slightly unusual for them, they would have a much greater probability or chance of lying outside the conventional reference interval. Generally, laboratories would flag these results as unusual in some way. Many clinicians would consider such a result as sig-

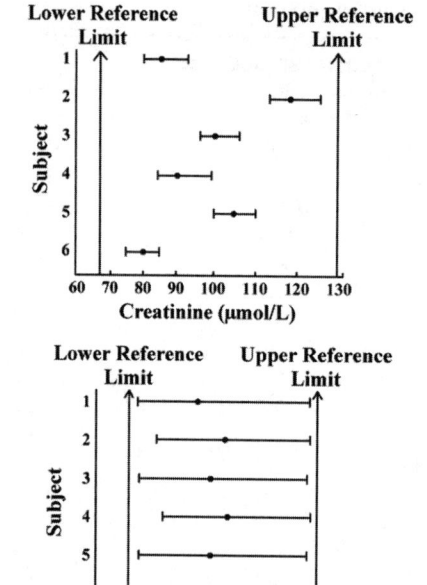

Figure 4.4 Means and Absolute Ranges for Serum Creatinine and Iron in Six Men

nificant and pay some attention to it and to other related test results, to results from other laboratory disciplines, and to results of other investigations and examinations so as to decide the action to be taken.

THE INDEX OF INDIVIDUALITY

The difference between creatinine and iron is the individuality of the analyte. Creatinine has marked individuality, while iron has little individuality. Creatinine has a low index of individuality, while iron has a high index. For creatinine, conventional population-based reference values are of limited utility in detecting unusual results in most individuals. In contrast, population-based reference values are of benefit for iron because unusual values for all individuals will lie outside the reference limits. Creatinine in the population as a whole has an index of individuality of 0.33, whereas the index for iron is 1.14.

In his groundbreaking studies on individuality, Harris explored the consequences of this difference in detail. When II is low, particularly when it is less than 0.6, the dispersion of values for any individual will span only a small part of the reference interval. Reference values will be of little utility, particularly for deciding whether change has occurred. For analytes with marked individuality, comparison with previous values has many advantages.

In contrast, when II is high, particularly when it is higher than 1.4, the distribution of values from a single individual will cover much of the entire distribution of the reference interval derived from reference subjects. Thus, conventional reference values will be of significant value in many clinical settings. Very few analytes have an index of individuality greater than 1.4; in many, II is less than 0.6. Therefore, population-based reference values are generally not good guides as to whether changes have occurred.

We can increase II by stratification, thereby making reference values more useful. Take urine creatinine, for example. The within-subject and between-subject components of variation are shown in Table 4.2, and the daily output for urine creatinine in the subjects studied (8 women, 7 men) is shown in Figure 4.5. As expected, women generally have lower urine creatinine output than men.

For the whole group, II is 0.46, making reference values not particularly useful. For women and for men separately, II are 1.42 and 1.83 respectively, making reference values very useful. Stratification according to sex has vastly increased the utility of conventional population-based reference values. Since most analytes have low II, stratification, which increases II, should be considered when we are developing reference values.

INDICES OF INDIVIDUALITY AND DIAGNOSIS

From the point of view of biological variation, the ideal test would have a small within-subject biological variation. The RCV would be small, which would make modest changes in an individual's serial results of real significance. Moreover, the ideal test would have low individuality, which would make population-based reference values useful in many clinical settings, eliminating any need for stratification.

However, most tests have marked individuality. This not only affects the low utility of reference values in the monitoring situation, but also has consequences in diagnosis, case finding, and screening, where clinicians often use reference values to help them interpret results.

Some years ago, we studied the biological variation of immunoglobulin assays. It had been suggested that if we measured concentrations of serum IgG, IgA, and IgM, and κ and λ chains simply and directly on automated systems, we

Table 4.2 CV_I, CV_G, and II for Urine Creatinine in Women and Men

Group	Within-subject variation (%)	Between-subject variation (%)	II
Whole	13.0	28.2	0.46
Women	15.7	11.0	1.42
Men	11.0	6.0	1.83

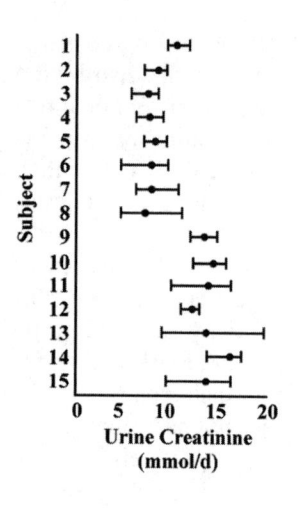

Figure 4.5 Means and Absolute Ranges for Urine Creatinine in 8 Women (Subjects 1–8) and 7 Men (Subjects 9–15). (Adapted with permission from Figure 3 in Gowans EMS, Fraser CG. Biological variation of serum and urine creatinine and creatinine clearance: ramifications for interpretation of results and patient care. Ann Clin Biochem 1988;25:259–263.)

could identify disorders of immunoglobulin production very easily. This would have many advantages, because the approach then in vogue involved protein electrophoresis and then immunoelectrophoresis or immunofixation, both of which are time consuming and require considerable expertise to perform and interpret.

The idea was that we could calculate the heavy chain to light chain ratio—

$$(IgG + IgA + IgM)/(\kappa + \lambda)$$

—and the κ/λ ratio to see if abnormal proteins were present and then, through quantitation, decide the type of protein present. The means and ranges of the κ/λ ratio are shown in Figure 4.6 for 12 healthy subjects.

This ratio has extreme individuality. In fact, it is the most individual test that we have seen to date. All of the immunoglobulins and the ratios have low II (so do other plasma proteins) as shown in Table 4.3. This means that individuals can have very unusual values, but the values will still lie within the reference intervals, prompting neither laboratories nor clinicians to investigate further. We can pick up small bands on electrophoresis but we cannot detect these using the proposed technique, even though it is convenient and easy to automate.

Other studies showed that this technique picked up only 71–95% of identified paraproteins. Our simple studies on biological variation clearly demonstrated the reason for this lack of clinical sensitivity—individuals can have values that are unusual values for them but still within the reference limits. Individuality—the usual finding—provides a logical explanation for why laboratory tests are not good at detecting latent, early, or pre-symptomatic disease, and why using a battery of laboratory tests in screening or case finding has never really been shown to be of significant benefit.

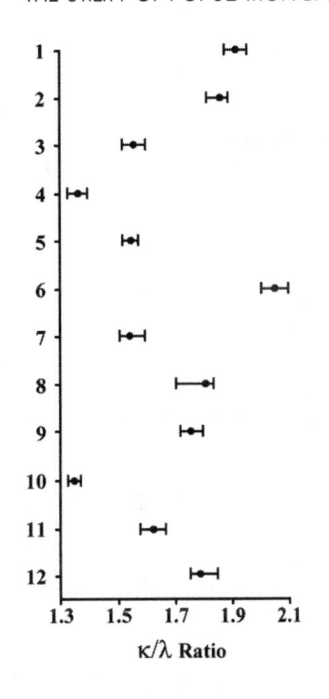

THE UTILITY OF POPULATION-BASED REFERENCE VALUES

Figure 4.6 Means and Absolute Ranges for the κ/λ Ratio in 12 Healthy Subjects. (*From Figure 2 in Ford RP, Mitchell PEG, Fraser CG. Desirable performance characteristics and clinical utility of immunoglobulin and light-chain assays derived from data on biological variation. Clin Chem 1988;34:1733–1736, used with permission.*)

INDIVIDUALITY IN HEMATOLOGY

While most of the examples used in this book are from clinical biochemistry (reflecting my background), we should note that individuality is usual in all facets of laboratory medicine.

We have studied the biological variation of the full blood count in apparently healthy elderly people. Others have studied the biological variation of a variety of analytes assayed in hematology laboratories.

Figure 4.7 shows the means and absolute ranges for hemoglobin and hematocrit. Subjects 1–12 are men and subjects 13–24 are women.

Visual interpretation shows the following.

- Hemoglobin and hematocrit have marked individuality and will have low indices of individuality.
- In general, men have higher values than women, so that stratification according to sex would increase the index of individuality and make conventional reference values more useful.
- Trying to assess changes in any individual using reference limits will be less useful than using RCV.
- Even on stratification, individuality is high, so that people could have values that would be very unusual for them but still lie within conventional reference values. In this case, performing hematology tests for screening or case finding will not have the desired 100% sensitivity, and false negatives will be common.

Table 4.3 CV$_I$, CV$_G$, and II for Immunoglobulins, Light Chains and Derived Ratios

Analyte	Within-subject variation (%)	Between-subject variation (%)	II
IgG	4.4	13.0	0.34
IgA	5.0	35.0	0.14
IgM	5.9	48.5	0.12
κ	4.8	15.3	0.31
λ	4.8	17.3	0.28
κ/λ ratio	0.7	12.1	0.06
H/L ratio	4.2	4.8	0.87

In Table 4.4, we provide average biological variation data for components of the full blood count, taken from the latest database.

Individuality is marked for all hematological analytes and indices studied to date. It is also marked for tests, including cell counts, performed in clinical immunology laboratories.

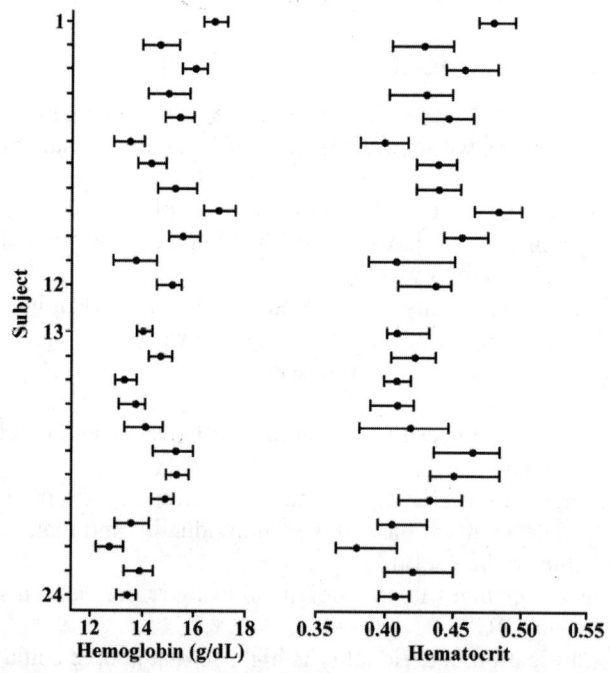

Figure 4.7 Means and Absolute Ranges for Hemoglobin and Hematocrit in Elderly Subjects. (*From Figure 1 in Fraser CG, et al. Biological variation of common hematologic laboratory quantities in the elderly. Am J Clin Pathol 1989;92:465-470. Used with permission.*)

THE UTILITY OF POPULATION-BASED REFERENCE VALUES

Table 4.4 CV_I, CV_G, and II for Commonly Assayed Hematology Analytes and Ratios

Analyte	Within-subject variation (%)	Between-subject variation (%)	II
Hemoglobin	2.8	6.6	0.42
Hematocrit	2.8	6.6	0.42
MCV	1.3	4.8	0.27
MCH	1.6	5.2	0.31
MCHC	1.7	2.8	0.61
Erythrocytes	3.2	6.1	0.52
Leukocytes	10.4	27.8	0.37
Platelets	9.1	21.9	0.42

INDIVIDUALITY AND REPEATING TESTS IN THE HEALTHY

Conventional population-based reference intervals encompass 95% of the population. This is partly by convention but is also recommended by the IFCC and other national and international authorities.

This means that 5% of the reference population will have values outside the reference limits—2.5% will have values higher than the upper reference limit, and 2.5% will have values lower than the lower reference limit.

These people are not diseased or ill. They are simply different from most of the distribution of the reference population. Many users of laboratory services know that an unusual result does not always signify a clinical problem. However, other users of laboratory results often repeat an "abnormal" value, without giving it a great deal of thought.

Very often, the repeat result falls within the reference interval and the person who requested the test is satisfied—possibly thinking, "That laboratory made a mistake," or, more likely, not pondering the reason for the change at all.

When an analyte has marked individuality, the values found for each individual span only a small part of the range of values observed in the reference population. Naturally, some will have homeostatic setting points close to the reference limits as shown in the middle diagram of Figure 4.8. These setting points may be outside the limits or inside the limits.

Due to inherent sources of variation, however, when we repeat a test just because it is a little outside a limit (a "borderline" result), the repeat test result may well be inside the limit. Thankfully, we do not repeat test results inside the reference interval too frequently because people with homeostatic setting points just inside the limits might become unusual—due simply to inherent sources of variation—and make the clinician think that a new disease is present.

When an analyte has little individuality, all people will have homeostatic setting points within the reference interval and their values will span nearly the whole distribution as shown in the bottom diagram of Figure 4.8. The chance of finding

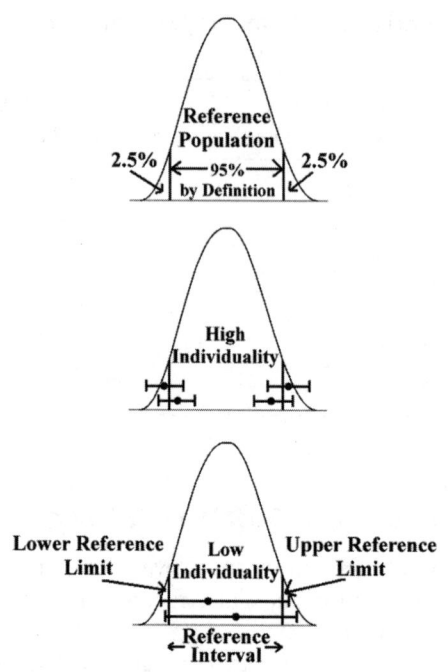

Figure 4.8 Consequences of Repeating a Test Just Outside Reference Limits in Healthy People

an unusual result in the reference population is, of course, 5%. However, if we repeat an unusual result (particularly a borderline result) because it is unusual, the repeat result will most likely lie within the reference interval. However, if we repeat results inside the reference interval for some reason, chances are these would be usual as well, with a low probability of being unusual.

Therefore, regardless of whether an analyte has high or low individuality, if we repeat tests on the 5% of the population whose results fall outside reference values, the repeat results will most likely lie within the reference interval. In case finding and screening, this is advantageous because we will not unnecessarily investigate people whose unusual result becomes usual on repeat.

INDIVIDUALITY AND REPEATING TESTS IN THE ILL

When we repeat unusual tests results found in healthy people (knowing that such results are found in 5% of the population by definition), generally the repeat result is within the reference interval. This holds true regardless of the individuality of the analyte. The result becomes usual because most values lie within the reference interval and individuals "regress towards their mean."

Now consider what happens when an individual becomes diseased. The upper diagram of Figure 4.9 shows an analyte with high individuality measured in just

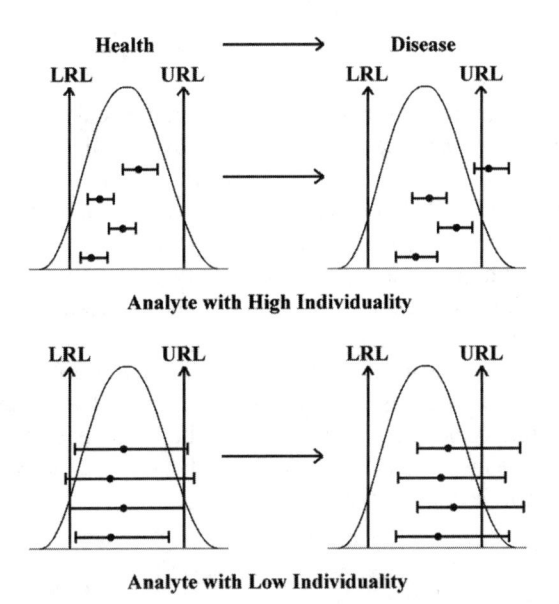

THE UTILITY OF POPULATION-BASED REFERENCE VALUES 113

Figure 4.9 Influence of Individuality on Repeating Tests in Ill People

four people with rather different homeostatic setting points. In this example, when people become diseased, their analyte values increase. The means and ranges for all four people also increase.

However, because of individuality, most values still lie within the reference interval nearly all of the time. The few people who have homeostatic setting points close to the reference limit are thus actually fortunate in this setting. For example, the setting point of the topmost subject in the upper diagram was near the upper reference limit and so, when the values start to change, reflecting progression of disease, the homeostatic setting point moves higher than the upper reference limit.

If we repeat this test to "confirm" the unusual finding, then it will probably still be unusual, and repeating "markedly unusual" (not just borderline) test results may just waste resources.

When analytes have marked individuality, the chance of disease being picked up does increase as disease progresses and the homeostatic setting point becomes closer to, and then exceeds, the reference limit.

The lower diagram of Figure 4.9 shows an analyte with low individuality. In this example as well, when people become diseased, their analyte values, including means and ranges, increase. But, because of the lack of individuality, most values still lie within the reference interval nearly all of the time, even though all have values greater than the reference limit some of the time.

Thus, in contrast to analytes with marked individuality—when few become unusual, but when they do it is significant—here all become unusual for some of the time. Everyone has a chance of being picked up as diseased, but only a small chance. As disease progresses and the analyte becomes more and more unusual, the probability of detection will increase—and will increase more or less the same for everybody.

If we repeat the unusual tests to "confirm" the unusual finding for an analyte with low individuality, then the repeat result is likely to lie within reference limits—simply because of the shape of the Gaussian distribution.

Only when the values become really unusual are they likely to remain unusual on repeat. In consequence, repetition will give us the wrong answer unless disease is severe. Repeating an unusual result is not a very productive procedure diagnostically, although it does have some benefits in that the mean of replicate analyses has a smaller dispersion than that of a singleton estimate.

Therefore, repeating test results to "confirm" an unusual result in a diseased population is either not very useful because the test result will stay unusual when individuality is high, or not very useful because the repeat has a high chance of being normal when the individuality is low. Regardless of the index of individuality, test repetition is not very effective.

SUMMARY

We covered the following key points in this chapter.

- We use clinical laboratory test results for many purposes.
- We commonly use population-based reference values as guides to interpreting test results.
- The words "normal" and "abnormal" have many meanings and are inappropriate for values given by laboratories to aid in interpretation.
- The reference value concept and terminology should be used; reference individuals make up a reference population from whom can be selected a reference sample group on whom are determined reference values on which is observed a reference distribution on which is determined reference limits that define a reference interval.
- International recommendations do not support using patient data to derive reference values.
- Reference intervals can be selected *a priori* by setting criteria for inclusion or exclusion before doing the tests, or *a posteriori* by application of the criteria only after the tests have been done on a larger group.
- We ought to minimize sources of pre-analytical variation and conduct analysis over a number of well-controlled quality analytical runs.
- The reference interval should be the non-parametric central 95%.
- Reference values are affected by endogenous, exogenous, genetic/ethnic, laboratory, and statistical factors, and may require stratification according to factors such as age and sex.
- Simple techniques remain available to assist in deciding whether stratification of reference values into sub-groups is necessary.

- Generating and applying conventional reference values have many problems; we recommend a pragmatic approach to developing them in laboratories.
- The variation of most individuals over time is less than the dispersion of the reference interval, and within-subject biological variation is generally less than between-subject variation.
- The ratio of within-subject biological variation to between-subject biological variation is called the index of individuality, and the index of individuality is low for most analytes; this means marked individuality is present.
- Population-based reference values are of low utility for monitoring people for analytes with a low index of individuality, particularly when < 0.6.
- Population-based reference values are of general use for analytes with a high index of individuality, particularly when > 1.4.
- Stratification increases the index of individuality and makes reference values more useful for monitoring, as well as for diagnosis.
- Marked individuality shows why laboratory tests are poor for case finding and screening, because individuals can have values that are very unusual for them but that still lie within the conventional reference interval.
- Even in significant disease, only those with homeostatic setting points close to the reference limit will be detected if the analyte has high individuality; repeating this unusual result will probably give the same result.
- If an analyte has little individuality, all individuals stand a similar chance of being detected as the disease progresses: Repeat values will have some chance of lying within the reference values simply due to inherent variation.

FURTHER READING

1. Faulkner WR, Meites S. Geriatric clinical chemistry reference values. Washington, DC: AACC, 1993. (*Source of reference values for the elderly.*)
2. Harris EK, Boyd JC. On dividing reference data into sub-groups to produce separate reference ranges. Clin Chem 1990;36:265–270.
3. Harris EK. Statistical aspects of reference values in clinical pathology. Prog Clin Pathol 1981;8:45–66.
4. Henny J, Peticlerc C, Fuentes-Arderiu X, et al. Need for revisiting the concept of reference values. Clin Chem Lab Med 2000;38:589–595.
5. Hyltoft Petersen P, Fraser CG, Sandberg S, Goldschmidt H. The index of individuality is often a misinterpreted quantity characteristic. Clin Chem Lab Med 1999;37:655–661.
6. Meites S. Pediatric clinical chemistry reference (normal) values, 3rd ed. Washington, DC: AACC, 1989. (*Source of reference values for children.*)
7. NCCLS C28-A: How to define and determine reference intervals in the clinical laboratory; approved guideline—2nd edition. Villanova, PA: National Committee for Clinical Laboratory Standards, 2000.
8. Siest G, Henry J, Schiale F, Young DS. Interpretation of clinical laboratory tests. Basle: Karger, 1985. (*Excellent for age and sex stratified values.*)
9. Sinton TJ, Crowley D, Bryant SJ. Reference values of calcium, phosphate, and alkaline phosphatase as derived on the basis of multi-analyzer profiles. Clin Chem 1986:32:76–79.

10. Solberg HE, Grassbeck R. Reference values. Adv Clin Chem 1989;27:1–79. *(A detailed review of all six IFCC Approved Guidelines.)*
11. Tietz NW. Clinical guide to laboratory tests. Philadelphia: Saunders, 1988. *(Good for rarely performed assays and very comprehensive regarding drugs, hematology, microbiology].*
12. Whitehead TP, Robinson D. Hale AC, Bailey AR. Clinical chemistry and haematology: adult reference values. London: BUPA Medical Research and Development Ltd., 1994.

Chapter 5
Other Uses of Data on Biological Variation

The *major* uses of data on the components of analytical and biological variation are: (1) setting quality specifications, (2) assessing the significance of changes in serial results from an individual, and (3) evaluating the utility of conventional population-based reference values. However, the data can also be used for a variety of other interesting purposes, as described below.

EPIDEMIOLOGY

We have examined the individuality of tests and the consequences of marked individuality in diagnosis, case finding, screening, and monitoring. Epidemiologists use very similar information, but in a slightly different way. They use the *reliability coefficient*

$$CV_G^2 / (CV_A^2 + CV_I^2 + CV_G^2)$$

or, the between-subject variance divided by the total variance. The reliability coefficient, usually called R, is numerically equal to the correlation coefficient of repeated measurements. It can be between 0 and 1 (the index of individuality can lie between and 0).

If R approached 1, then this would be the same as having a very low index of individuality (CV_A/CV_G would be low). The test would therefore have very marked individuality. Thus, individuals would vary little around their own homeostatic setting points, but the setting points of individuals would differ greatly from each other. In this situation, any single test result would be a relatively reasonable measure of that individual's homeostatic setting point. In addition, over time, test results would vary, but not throughout the entire reference interval.

In contrast, if R approached 0, then this would be the same as having a very high index of individuality. The test would therefore show little individuality. Individuals would vary considerably around their own homeostatic setting points, the setting points of different individuals would be similar, and any single test result would be only a crude measure of the homeostatic setting point of that individual. In addition, over time, test results for each individual would vary considerably through a large fraction of the reference interval.

Thus R, used in epidemiology and sometimes calculated in the literature of laboratory medicine, gives similar information to the more widely used index of individuality. As we have seen, individuality is usual for analytes in blood: R will usually be high, too.

NUMBER OF SAMPLES NEEDED

In usual clinical practice, we take only one sample. However, even if analytes have a high index of individuality and a single test result does reasonably estimate the individual's homeostatic setpoint, the result does have intrinsic analytical and within-subject biological variation. As mentioned earlier, variation can be reduced by multiple sampling or multiple analyses, and the variation is made smaller by the reciprocal of the square root of the number of replicates performed. In numerical terms, the relevant SD or CV is simply multiplied by $1/n^{1/2}$. Performing assays of a single sample in duplicate not only catches analytical blunders, but also reduces the analytical component of the variation (CV_A) by $1/2^{1/2}$, i.e., to 70% of the singleton value.

To calculate how many samples we need to ensure that our estimate of the homeostatic setting point is within a certain percentage of the true value with a stated probability, we use a simple rearrangement of the usual standard error of the mean formula:

$$n = [Z*(CV_A^2 + CV_I^2)^{1/2}/D]^2$$

where Z is the Z-score appropriate for the probability. This is usually 1.96 for $p<0.05$, and D is the desired percentage of closeness to the homeostatic setting point.

Take serum cholesterol as an example. Authors of many of the national recommendations regarding cholesterol appreciate that a single numerical test result has variation, and therefore recommend that two samples be taken. As discussed earlier, the within-subject biological variation was 6%. If the laboratory achieved the maximum precision recommended by the U.S. National Cholesterol Education Program, CV_A would be 3%.

Thus, if we wanted our estimate to be within 10% of the true homeostatic setting point with 95% probability,

$$n = [Z*(CV_A^2 + CV_I^2)^{1/2}/D]^2$$
$$n = [1.96*(3^2 + 6^2)^{1/2}/10]^2$$
$$n = 2.$$

Lowering the probability (lowering the value of Z) or widening the allowance for the window of acceptability (increasing D) will decrease the number of samples required.

Analytical random variation, or precision, plays a role here as well. If the precision were $CV_A = 1\%$, then a single sample would actually suffice, but if CV_A were 10% as might be experienced with some point-of-care testing systems, then an unrealistic 5 samples would be needed. This deepens the rationale for achieving low precision in everyday practice.

OTHER USES OF DATA ON BIOLOGICAL VARIATION

REPORTING RESULTS

We sometimes report laboratory test results in different ways. Take, for example, the analysis of 24-hour urine collections. Usually, we measure the volume or weight, and then analyze an aliquot. We generate the assay result in terms of concentration (units per liter). Then, we calculate the output (units per day). The question then arises as to which result we should report.

From the viewpoint of biological variation, the ideal analyte has low within-subject biological variation so that one measurement will give a good estimate of the true value for that individual. Moreover, this would make for good monitoring because the RCV would be low. In addition, the ideal test would have no heterogeneity of within-subject variation so that the simple general formulae in this book would hold for everyone. Moreover, the index of individuality would be high so that conventional population-based reference values would be highly useful in diagnosis and case finding. The best way to report results is in such a way that the data most closely approaches these ideal characteristics.

A number of studies have investigated components of biological variation in commonly analyzed urine analytes. As Appendix 1 shows, reporting results as output—units per day—has advantages. For example, in our own extensive studies on creatinine, CV_I is 23.8% in terms of concentration, and 13.0% in terms of output.

SELECTING THE BEST SAMPLE TO COLLECT

We can sometimes collect different samples for the same clinical purpose. Considering the ideal test from the viewpoint of biological variation might assist us in deciding which sample to collect. For instance, measurement of low concentrations of albumin in urine (microalbumin) is very useful in diabetes care. We could, however, collect early morning urine samples, random spot samples when the diabetic patient appears at a health care site, or 24-hour urine samples, with results expressed as either concentration or output. Some of our data are shown in Table 5.1.

The first morning urine might be the best sample to collect, because it had the smallest CV_I. The fact that the first morning sample data for men and women were very similar, and individuals all had similar CV_I, supports this recommendation.

Because creatinine output remains "constant," it often appears as though every analyte should be "corrected for creatinine" (expressed as a ratio of unit per unit of creatinine). The NCCLS, in its document on reference values, states that creatinine ought to be checked to ensure the validity of 24-hour urine collections.

However, urine creatinine has significant inherent variation itself. Expressing albumin in urine as mg/mmol creatinine has significant disadvantages because it does not generally reduce CV_I, does not make men more like women, and does not make everyone's CV_I the same. This, however, is not true for all analytes, therefore we should carefully consider the benefits of expressing analytes as units per unit of creatinine, and not just accept it as dogma.

Table 5.1 Biological Variation for Microalbumin for Men and Women Combined

Sample type	CV$_I$ (%)	CV$_G$ (%)
First morning	36	35
Random spot	86	61
24 hour—as concentration	61	53
24 hour—as output	70	55

CHOOSING THE BEST TEST

We sometimes use different tests for the same or similar clinical purposes. Considering the ideal test keeping biological variation in mind may prove helpful if we are only going to perform the single most appropriate test, as opposed to doing more than one test. Sometimes there is little reason to choose one test over another—for example, we studied serum amylase and lipase in healthy subjects, and could not see any particular advantages or disadvantages in either.

We have, however, investigated some cardiac markers, and the RCV for CK-MB mass was very much smaller than for CK-MB activity, suggesting that the former would be better in monitoring change than the latter.

Keevil and colleagues have provided an excellent example. They studied two tests of renal impairment: a newer one that reports serum cystatin C, and a traditional test that measures serum creatinine. The components of biological variation and indices of individuality are shown in Table 5.2. These two tests have quite different characteristics. Creatinine is very individual, cystatin C is less so. The index of individuality for creatinine is 0.27 and for cystatin C it is 1.64, showing that population-based reference values would be more useful for cystatin C than for creatinine. Furthermore, the RCV for creatinine is much less than for cystatin C (since both analytical precision and within-subject biological variation are smaller). Thus, cystatin C might be more useful in diagnosis and creatinine more helpful in monitoring. This small study provides interesting information that could be obtained only from much larger clinical studies.

Table 5.2 Biological Variation for Cystatin C and Creatinine

Analyte	CV$_I$ (%)	CV$_G$ (%)	II
Cystatin C	13.3	8.1	1.64
Creatinine	4.9	18.2	0.27

OTHER USES OF DATA ON BIOLOGICAL VARIATION

Before cystatin C became available as a renal function test, we had examined creatinine clearance and serum creatinine for the same purpose, and we reached similar conclusions: serum creatinine had low CV_I and a low RCV that made it good for monitoring. However, it had a low index of individuality, even when stratified for men and women separately, suggesting that reference values would be of little value in detecting minor renal impairment.

Creatinine clearance, however, had the opposite characteristics and had a single set of useful reference values making it, theoretically at least, more suitable for investigating small loss of renal function. Again, a small study done on a few samples from a cohort of healthy people gave a great deal of information that one might think could only be obtained through complex and lengthy clinical investigation.

METHOD DEVELOPMENT AND EVALUATION

Introducing new test procedures remains an ongoing task for most clinical laboratories. Some years ago, Zweig and Robertson suggested that introducing a new procedure should be similar to the structured evolution of a new drug through phase trials. They suggested the following phases.

- Phase I: Analytical investigation—assessment of reliability and practicability characteristics
- Phase II: Overlap investigation—generation of reference values and assessment of values in health and disease
- Phase III: Clinical investigation—evaluation of sensitivity, specificity, and predictive value
- Phase IV: Outcome investigation—assessment of whether individuals gain advantage
- Phase V: Utility investigation—a cost-benefit analysis with respect to individuals and the community at large

One aspect not covered in these guidelines is the need to generate and apply data on biological variation early in the evolution of any new test procedure. We can generate data on the components of biological variation from very simple experiments involving duplicate analysis of a small number of samples from a small cohort of healthy individuals. This allows us to set quality specifications (needed for the Phase 1 investigation), calculate the significance of changes in an individual (unobtainable by other methods and not covered in the scheme above), and assess the utility of population-based reference values (as generated in the Phase II investigation).

Moreover, the biological variation data may yield other useful insights into the test and into how it compares with other procedures. Generating and applying data on biological variation are essential prerequisites for introducing new procedures.

SUMMARY

This chapter covered the following important points.

- The easy to calculate reliability coefficient used in epidemiology gives the same information as the index of individuality.
- We can calculate the number of samples needed to obtain an estimate of the homeostatic setting point within a certain percentage with a stated probability by using the Z-score, the percentage, and estimates of the precision and within-subject biological variation.
- Data on biological variation data may be of value in deciding the best way to report test results and the best sample to collect.
- When considering different test procedures for use in one clinical setting, we should take into account the data on biological variation.
- Generating and applying data on biological variation data are essential prerequisites in the evolution of any new test procedure.

FURTHER READING

1. Fraser CG. Data on biological variation; essential prerequisites for introducing new procedures. Clin Chem 1994;40:1671–1673.
2. Fraser CG, Harris EK. Generation and application of data on biological variation in clinical chemistry. Crit Rev Clin Lab Sci 1989;27:409–437.
3. Keevil BG, Kilpatrick ES, Nichols SP, Maylor PW. Biological variation of cystatin C: implications for the assessment of glomerular filtration rate. Clin Chem 1998:44:1535–1539.
4. Zweig MH, Robertson EA. Why we need better test evaluations. Clin Chem 1982;28:1272–1276.6

Chapter 6
Next Steps

Interested readers may want to take their knowledge of biological variation to a higher level. In this chapter, I suggest articles that are most appropriate as starting points for further study, and recommend immediate steps you can take to deepen your knowledge and experience of this subject. I also give some ideas on how you can use your new knowledge in your own laboratory.

You may wish to read Appendix 1 and the most recent database (available at www.westgard.com/guest17.htm), which illustrate the magnitude of biological variation. You will notice the following key points.

- Analytes with very good homeostatic control mechanisms—such as sodium, chloride, and calcium—have very low within-subject biological variation.
- Analytes with looser homeostatic control mechanisms—such as triglycerides, CK, and urea—have high within-subject biological variation.
- Analytes in urine tend to have high within-subject biological variation.
- The within-subject biological variation of cellular elements is related to the half-life of the cell type.
- High individuality (within-subject < between-subject variation) is usual. (In my opinion, this casts real doubt on whether we should spend significant time, effort, and resources in developing population-based reference values.)
- Most of the recommendations for desirable quality based on biology do not seem too difficult to attain with current methodology and technology.

REVIEW, NEXT STEPS, AND FURTHER READING

Chapter 1   In Chapter 1, The Nature of Biological Variation, we learned the following.

- Pre-analytical variation is an important source of test result variation.
- If we want to make the uncertainty of any test result smaller, the change in serial results from an individual as significant as possible, and the reference intervals as narrow as possible, we must reduce some sources of variation.
- It is impossible to reduce biological variation unless replicate samples are taken at the same time (to reduce variation by $1/n^{1/2}$), which would be totally impractical as well as costly.
- It may be difficult to reduce precision. Consequently, we should pay considerable attention to pre-analytical sources of variation.

Next steps might include the following.

- Revisit guidelines for patient preparation.
- Investigate guidelines for the collection of samples by phlebotomists.
- Update these guidelines if required.
- Institute training if needed.
- Revise material in User Guides and Laboratory Handbooks to educate and assist all users in collecting appropriate samples at suitable times in correct containers with sound handling techniques.
- Investigate procedures for sample delivery and handling in the laboratory, and ensure that process times, centrifugation techniques, and storage conditions are standardized so as to minimize variation.
- Develop and/or update the laboratory standard operating procedures relevant to pre-analytical processes and insist that all adhere to these procedures.

Recommended reading for Chapter 1

Fraser CG, Harris EK. Generation and application of data on biological variation in clinical chemistry. Crit Rev Clin Lab Sci 1989;27:409–437.

Chapter 2   In Chapter 2, Quality Specifications, we investigated the hierarchy of models available for setting quality specifications, particularly for precision, bias, and total error allowable, and we learned the following.

- Desirable, minimum, and optimum quality specifications are probably best derived from within-subject and between-subject biological variation.
- There are other useful ways to set quality specifications such as those laid down in CLIA'88.
- All models in the hierarchy could be used when objectively appropriate.

Next steps might include institution of quality planning. This involves the following.

- Investigate Appendix 2, which provides the quality specifications based on biology for precision, bias, and total error allowable.
- Assess your own precision at clinically important levels of analyte from your internal quality control program and your own bias from either PT or EQAS or peer review QC program, or from the original evaluation of the methodology (comparison of methods and recovery experiments).
- Then decide how many internal quality control samples to use per run and decide the QC rules to apply for acceptance and rejection.

This is easier to write down than it is to practice, particularly if we want to be very interactive with our clinical users and report results before the "run" has

NEXT STEPS

ended, and if we have many analytes on a single analytical system or combination of workstations run as a single system, all requiring different rules.

A number of available tools can help in quality planning, but a thorough discussion of these tools does not fall within the scope of this book. Interested readers who do not have the time or resources to implement full quality planning may consider the following pragmatic approach.

- Obtain, as above, the quality specifications based on biology for precision, bias, and total error allowable.
- Assess your own precision and bias and calculate your total error (perhaps concentrating on precision in the first instance since this contributes most to test result variation).
- Then rank the analytes in order of compliance with the quality specifications using the modified medical decision chart shown in Chapter 2, or simply divide the analytes into two groups: those that attain the specifications to any extent, and those that do not.
- Calculate critical systematic error (ΔSE_c) to assess the number of SD that the mean can shift before each control will exceed quality specifications. Concentrate on methods with low ΔSE_c.
- Concentrate attention and resources on improving methods for the analytes that least satisfactorily meet the quality specifications.
- Institute an improved internal quality control program for those analytes, one that ensures that the problems with the method are detected and that unsatisfactory results are not reported.

Recommended reading for Chapter 2

Fraser CG, Hyltoft Petersen P. Analytical performance characteristics should be judged against objective quality specifications. Clin Chem 1999;45:321–3.

Hyltoft Petersen P, Fraser CG, Kallner A, Kenny D, eds. Strategies to set global analytical quality specifications in laboratory medicine. Scand J Clin Invest 199;57:475–585.

Chapter 3 In Chapter 3, Changes in Serial Results, we learned the following.

- Most test results are used for monitoring.
- Both conventional population-based reference values and fixed limits have deficiencies for clinical monitoring.
- Comparing serial results is simple using RCV.

Next steps might include investigation of RCV in your own laboratory by doing the following.

- Obtain the most recent data available at Appendix 1 for within-subject biological variation.
- Assess your own precision at clinically important levels of each analyte from the internal quality control program.
- Calculate RCV at 95% probability (and 99% probability if deeply interested) using the simple formula

$$\mathrm{RCV} = 2^{1/2} * Z * [CV_A^2 + CV_I^2]^{1/2}$$

—where Z is 1.96 (or 2.58).

You can use RCV to flag patient reports with significant and/or highly significant change indicators, which would be a big step for many laboratories. You may need to talk with the local software expert or the LIMS vendor to facilitate this. Moreover, further education of both laboratory staff and users of test results would be required.

Readers who do not wish to do this, or do not have the time or resources to implement the concepts in full, could consider at least calculating the RCV for common analytes for which the laboratory already has delta-checking in place. You could then assess whether the delta-check values were actually doing the intended function of detecting changes greater than expected from analytical and biological variation. Using the 99% RCV might be of most value in this particular situation.

Recommended reading for Chapter 3

Queralto JM, Boyd JC, Harris EK. On the calculation of reference change values, with examples from a long-term study. Clin Chem 1993;39:1398–1403.

Chapter 4 In Chapter 4, The Utility of Population-Based Reference Values, we appraised the international recommendations on how reference values should be created by laboratories, and learned the following.

- These recommendations would take considerable time, expertise, and resources to implement fully.
- These reference values do have quite clear deficiencies, even if superbly generated.

Possible next steps include the following.

Readers could investigate the sources of their own in-house reference values, appraise the stratification or partitioning of their reference values, and then address any reference values that appeared of dubious or suspect origin.

The hierarchy of approaches detailed in Chapter 4 might help in this appraisal and regeneration, as follows:

NEXT STEPS

1. use IFCC recommendations exactly as published, then
2. use modified IFCC recommendations with laboratory or hospital staff, blood donors, or even patient data, then
3. use peer reviewed literature dealing with reference values, especially if same/similar methodology, then
4. use books on reference values, particularly for the young and old, and as aids in stratification, then
5. use other literature, and finally
6. use data from instrument or reagent manufacturers' literature.

You may wish to examine the goodness of the transfer of reference values from any source to the laboratory at regular intervals, and institute the 20-sample experiment recommended by NCCLS.

Recommended reading for Chapter 4

Solberg HE, Grassbeck R. Reference values. Adv Clin Chem 1989;27:1–79.

Chapter 5 In this Chapter, Other Uses of Data on Biological Variation, we looked at minor uses of data on biological variation, and learned the following.

- The reliability coefficient used in epidemiology is easy to calculate and gives the same information as the index of individuality.
- The formula that allows calculating the number of samples needed to obtain an estimate of the homeostatic setting point within a certain percentage with a stated probability uses the Z-score, the percentage, and estimates of the precision and within-subject biological variation.
- Data on biological variation may be valuable for deciding the best way to report test results and the best sample to collect.
- Considering data on biological variation when comparing different test procedures is potentially useful.
- Generating and applying data on biological variation is an essential prerequisite in the evolution of any new test procedure.

A FINAL NOTE

I hope that you have found biological variation to be as fascinating as I have, and that you will be stimulated to

- determine the components of biological variation of the many as yet unstudied analytes,
- explore new ways to use new and existing data on the components of biological variation,

- advance ways in which vignette studies can be used, along with probability and biological variation, to generate quality specifications directly from the opinions of clinician users,
- investigate objectively whether using reference change values is of real clinical benefit, and
- find simpler ways to introduce objective quality planning.

I also hope that you and your colleagues will "spread the word" and encourage professionals in fields other than clinical chemistry to learn the principles of generating and applying data on biological variation and to translate these into everyday practice.

Glossary

Analysis the procedural steps performed that enable the measurement of the amount of analyte in a specimen.
Analyte the constituent of the specimen to be measured.
A posteriori selection of subjects after performing measurements.
A priori selection of subjects before performing measurements.

B_A analytical bias—in calculations of total error, bias does not have a sign (positive or negative)—it is always positive.
Between-subject biological variation the difference between the homeostatic setting points of individuals.
Bias (of measurement) the difference between the results of measurement and the true value of the measured quantity; in practice, bias is the difference between the results and an estimate of the true value.

Case finding the opportunistic performance of a panel of investigations, usually including a range of clinical laboratory tests, when an individual presents to the health care system.
Circadian depending on the time of day.
Coefficient of variation (CV) a measure of relative precision calculated as a standard deviation (SD) divided by its mean and often multiplied by 100 and expressed as a percentage.
Critical Systematic Error, or $\Delta SE_c = [(TE_a - B_A)/CV_A - 1.65]$ a good indicator of our method performance, indicating, in a single statistic, the number of SD the mean can shift before exceeding the total error allowable quality specification.
CV_A analytical coefficient of variation.
CV_G between-subject biological coefficient of variation.
CV_I within-subject biological coefficient of variation.
CV_P pre-analytical coefficient of variation.
CV_T total coefficient of variation.

Delta-check a technique for quality assurance, using patient data, based on investigating the differences between two values on individuals that are greater than a previously set value.
Diagnosis a process that involves identifying disease by investigating the symptoms, usually including performance of a range of relevant clinical laboratory tests.

Endogenous   inherent in the individual.
EQAS   external quality assessment scheme—determination of laboratory measurement performance by means of inter-laboratory comparisons and evaluation of the appropriateness of results from individual laboratories according to recognized criteria.
Exogenous   able to be influenced by the individual.

False negatives   results from diseased people that are not unusual.
False positives   results from well people that are unusual.
Fixed limit   a value set by some means that is used as a criterion for clinical action or as a criterion for acceptability in PT or EQAS.
Flag   a symbol used to highlight a value.

Gaussian distribution   a symmetrical bell-shaped distribution with a defined mathematical function.

Hierarchy   a list of approaches with the most closely approaching the ideal at the top and the least good at the bottom.
Highly significant   usually applied to 99% probability ($p < 0.01$).
Homeostatic model   a model for time series analysis that assumes that the analyte behaves in a random manner and that the individual has a homeostatic setting point.
Homeostatic setting point   the value around which the results from an individual vary over time.

Index of individuality   the ratio of within-subject to between-subject variation, most often calculated as the simple ratio of the biological components of variation (CV_I/CV_G).
Individuality   the property of being characteristic of a particular person.

LIMS   laboratory information management systems.

Monitoring   following over time which usually involves reviewing laboratory test results over time to assess change in serial results.

Negative bias   bias that results in methods giving values lower than the true values.
Non-parametric   statistical techniques that do not make any assumptions about the type of underlying distribution.
Normal   a term with so many meanings that its use is not recommended.
Nycthemeral   depending on sleep/wake cycle.

Outcome   a change in health status and quality of life resulting from a specified intervention.

GLOSSARY

Parametric statistical techniques that assume that the distribution of results is Gaussian.
Partition to divide the group into sub-groups.
Performance characteristics the attributes of an analytical method.
Positive bias bias that results in methods giving values higher than the true values.
Practicability performance characteristics the attributes of an analytical method that relate to the execution of the procedure.
Pre-analytical variation the variation arising from sources before the analytical part of the process of obtaining a test result.
Precision (of measurement) the closeness of agreement between independent results of measurements obtained under stipulated conditions.
PT proficiency testing—determination of laboratory measurement performance by means of inter-laboratory comparisons and evaluation of the appropriateness of results from individual laboratories according to recognized criteria.

Quality assurance the practice that that encompasses all procedures and activities directed towards ensuring that a specified quality is achieved and maintained.
Quality control set of procedures undertaken for the continuous monitoring of operation and the results of measurements.
Quality laboratory practice the techniques adopted in the laboratory to ensure that a specified quality is achieved and maintained.
Quality planning procedures undertaken using quality specifications to decide the number of quality control samples analyzed and the rules used for acceptance or rejection.
Quality specifications the level of performance required to ensure that a test fulfills its stated or implied purpose—also known as *quality goals, quality standards, analytical goals, analytical performance goals*.

Random variation variation that is not in a constant direction.
Random walk model a model for time series analysis that assumes that the analyte behaves randomly but not around a homeostatic setting point.
Reference change value the value that must be exceeded before a change in serial results from an individual is significant at a pre-set probability.
Reference distribution the statistical dispersion of the reference values.
Reference individual an individual selected for comparison using defined criteria.
Reference interval the interval bound by the upper and lower reference limits.
Reference limit value that defines a pre-set fraction of the reference values less than or equal to the limit.
Reference population the population that consists of all possible reference individuals.
Reference sample group an adequate number of reference individuals taken to represent the reference population.

Reference values   the values obtained on reference individuals for an analyte.
Reliability coefficient   the ratio of between-subject to total variation—a measure of individuality.
Reliability performance characteristics   the attributes of an analytical method that relate to the scientific facets, including precision and bias.

Screening   the identification of unrecognized disease or defect.
SD   standard deviation.
SD$_A$ analytical standard deviation.
SD$_G$ between-subject biological standard deviation.
SD$_I$ within-subject biological standard deviation.
SD$_P$ pre-analytical standard deviation.
SD$_T$ total standard deviation.
Significant   usually applied to 95% probability ($p < 0.05$).
Standard deviation   the square root of the variance; an estimate of the dispersion of a series of measurements on the same analyte.
State of the art   the level of performance achieved with currently available methodology and technology.
Stratify   to divide the group into sub-groups (same as *partition*).
Systematic variation   variation in a constant direction.

Time series analysis   techniques for assessing the significance of changes in series of results from an individual.
Total error (TE)   usually defined as bias $+$ 1.65 $*$ precision;—but other formulae exist; the general formula is TE = bias $+$ n $*$ precision.
Total error allowable (TE$_a$)   the quality specification for total error.
Transferability   the ability to validly use one set of data (reference values) to the laboratory from another source.
Transformation   modification of a set of data using a constant mathematical function.
True negatives   results from well people that are not unusual.
True positives   results from diseased people that are unusual.

Vignette   a short generalized clinical case history.

Within subject biological variation   the inherent biological variation around the homeostatic setting point.

Z-score   the *standard normal deviate*: the number of standard deviations appropriate to the selected probability.

Appendix 1
Data on Components of Biological Variation

The most recent and extensive listing of data on components of biological variation and many useful derived indices have been provided by Carmen Ricos and her group from Spain. The full list was published as

> Ricos C, Alvarez V, Cava F, Garcia-Lario JV, Hernandez A, Jimenez CV, Minchinela J, Perich C, Simon M. Current databases on biologic variation: pros, cons, and progress. Scand J Clin Lab Invest 1999;59:491–500.

The full database, together with the original references to the sources of the data, is available on the Internet at www.westgard.com/guest17.htm.

These three tables provide data on within-subject (CV_I) and between-subject (CV_G) components of biological variation for commonly assayed analytes in

1. serum or in whole blood (indicated in parentheses),
2. urine, and
3. hematology and hemostasis.

Additional data on very rarely done analytes do exist, so before you set out to determine data on analytes not included in these three tables, please go back to the original database to assess whether the data you seek already exist.

Table A1.1   Analytes in Serum or Whole Blood

Analyte	CV_I (%)	CV_G (%)
Alanine aminotransferase	24.3	41.6
Albumin	3.1	4.2
Aldosterone	29.4	40.1
Alkaline phosphatase	6.4	24.8
Amylase	9.5	29.8
Androstendione	11.5	51.1
Angiotensin converting enzyme (ACE)	12.5	27.7
Apolipoprotein-A1	6.5	13.4
Apolipoprotein-B	6.9	22.8
Ascorbic acid	25.8	22.9
Aspartate aminotransferase	11.9	17.9
β_2-microglobulin	5.9	15.5
β-carotene	36.0	39.0
Bilirubin (total)	25.6	30.5
Bilirubin (conjugated)	36.8	43.2
CA-125 antigen	13.6	46.5
CA-15.3 antigen	5.7	42.9

Table A1.1 Analytes in Serum or Whole Blood (*Continued*)

Analyte	CV_I (%)	CV_G (%)
CA-19.9 antigen	24.5	93.0
CA-549 antigen	9.1	33.4
Calcium	1.9	2.8
Carbohydrate deficient transferrin	7.1	38.7
Carcinoembryonic antigen (CEA)	9.3	55.6
Ceruloplasmin	5.7	11.1
Chloride	1.2	1.5
Cholesterol	6.0	15.2
Cholinesterase	7.0	10.4
C3 complement	5.2	15.6
C4 complement	8.9	33.4
Copper (serum)	4.9	13.6
Cortisol	20.9	45.6
C-peptide	9.3	13.3
C-reactive protein (CRP)	52.6	84.4
Creatine kinase	22.8	40.0
Creatine kinase MB, %	6.9	42.8
Creatine kinase MB, activity	19.7	24.3
Creatine kinase MB, mass	18.4	61.2
Creatinine	4.3	12.9
Dehydroepiandrosterone sulfate (DHEAS)	3.4	30.0
Estradiol	22.6	24.4
Ferritin	14.9	13.5
Follicle-stimulating hormone (FSH)	10.1	32.0
Free estradiol	22.8	—
Free testosterone	9.3	—
Free thyroxine (FT4)	7.6	12.2
Free tri-iodothyronine (FT3)	7.9	—
Fructosamine	3.4	5.9
γ-Glutamyltransferase	13.8	41.0
Globulins (total)	5.5	12.9
Glucose	6.5	7.7
Glycated albumin	5.2	10.3
Glycated total protein	0.9	11.6
Glycated hemoglobin (blood)	5.6	—
Haptoglobin	20.4	36.4
HDL-cholesterol	7.1	19.7
Hydroxybutyrate dehydrogenase	8.8	—
Homocysteine	7.7	29.9
Immunoglobulin A	5.0	36.8
Immunoglobulin G	4.5	16.5
Immunoglobulin M	5.9	47.3
Immunoglobulin κ-chain	4.8	15.3
Immunoglobulin λ-chain	4.8	18.0
Insulin	21.1	58.3

DATA ON COMPONENTS OF BIOLOGICAL VARIATION

Table A1.1   Analytes in Serum or Whole Blood (Continued)

Analyte	CV_I (%)	CV_G (%)
Iron	26.5	23.2
Lactate dehydrogenase 1	6.3	10.2
Lactate dehydrogenase 2	4.9	4.3
Lactate dehydrogenase 3	4.8	5.5
Lactate dehydrogenase 4	9.4	9.0
Lactate dehydrogenase 5	12.4	13.4
Lactate (blood)	27.2	16.7
Lactate dehydrogenase	6.6	14.7
LDL-cholesterol	8.3	25.7
Lipase	23.1	33.1
Lipoprotein (a)	8.5	85.8
Luteinizing hormone (LH)	14.5	27.8
Magnesium	3.6	6.4
Mucinous carcinoma-associated antigen (MCA)	10.1	39.3
Myoglobin	13.9	29.6
Osmolality	1.3	1.2
Osteocalcin	6.3	23.1
pCO_2 (blood)	4.8	5.3
pH (blood)	3.5	2.0
Phosphate	8.5	9.4
Phospholipids	6.5	11.1
Potassium	4.8	5.6
Prealbumin	10.9	19.1
Prolactin (men)	6.9	61.2
Proteins (total)	2.7	4.0
Prostate-specific antigen (PSA)	14.0	72.4
Pyruvate (blood)	15.2	13.0
Rheumatoid factor (RF)	8.5	24.5
Sex-hormone-binding globulin (SHBG)	12.1	42.7
Sodium	0.7	1.0
T3-uptake	4.5	4.5
Testosterone	8.8	21.3
Thyroglobulin	13.0	25.0
Tissue polypeptide antigen (TPA)	28.7	40.4
Tissue polypeptide specific antigen (TPS)	36.1	108.0
Thyrotropin (TSH)	19.7	27.2
Thyroxin binding globulin (TBG)	6.0	6.0
Thyroxine (total T4)	6.0	12.1
Transferrin	3.0	4.3
Triglycerides	21.0	37.2
Triiodothyronine (TT3)	8.7	14.4
Urate	8.6	17.2
Urea	12.3	18.3
VLDL-cholesterol	27.6	—
Zinc (serum)	9.3	9.4

Table A1.2 Analytes in Urine

Analyte	CV_I (%)	CV_G (%)
α_1-Microglobulin, concentration, overnight	33.0	58.0
α_2-Microglobulin output, overnight	29.0	32.0
Albumin, concentration, first morning	36.0	55.0
Aldosterone, concentration	32.6	39.0
Ammonia, output	24.7	27.3
Amylase activity, random	94.0	46.0
Calcium, concentration	27.5	36.6
Calcium, output	26.2	27.0
Catecholamines (total), concentration	24.0	32.0
C-Telopeptide type I collagen/creatinine, first urine	35.1	—
Creatinine, concentration	24.0	24.5
Creatinine, output	11.0	23.0
Deoxypyridinoline/creatinine	14.7	15.1
Estradiol	30.4	—
Free estradiol	38.6	—
Free testosterone	51.7	—
Hydroxyindolacetic acid, concentration, 24 h	20.3	33.2
Hydroxyproline/minute, night urine	36.1	38.8
Magnesium, concentration	45.4	37.4
Magnesium, output	38.3	37.6
N-Acetylglucosaminidase, concentration	52.5	33.5
N-Acetylglucosaminidase, output	42.4	18.2
Nitrogen, output	13.9	24.2
N-telopeptide type I collagen /creatinine, first urine	23.1	—
Oxalate, concentration	44.0	18.0
Oxalate, output	42.5	19.9
Phosphate, concentration	26.4	26.5
Phosphate, output	18.0	22.6
Potassium, concentration	27.1	23.2
Potassium, output	24.4	22.2
Protein, concentration	39.6	17.8
Protein, output	35.5	23.7
Sodium, concentration	24.0	26.8
Sodium, output	28.7	16.7
Urate, concentration	24.7	22.1
Urate, output	18.5	14.4
Urea, concentration	22.7	25.9
Urea, output	17.4	25.4
VMA, concentration	22.2	47.0

Concentration data are expressed as units/volume. Output data are expressed as units/24h

DATA ON COMPONENTS OF BIOLOGICAL VARIATION

Table A1.3 Analytes in Hematology and Hemostasis

Analyte	CV_I (%)	CV_G (%)
Hemoglobin	2.8	6.6
Hematocrit	2.8	6.4
Mean corpuscular hemoglobin	1.6	5.2
Mean corpuscular hemoglobin concentration	1.7	2.8
Mean corpuscular volume	1.3	4.8
Red cell distribution width	3.5	5.7
Erythrocyte count	3.2	6.1
Leukocyte count	10.9	19.6
Neutrophil, count	16.1	32.8
Monocyte count	17.8	49.8
Eosinophil count	21.0	76.4
Basophil count	28.0	54.8
Platelet count	9.1	21.9
Mean platelet volume	4.3	8.1
Platelet distribution width	2.8	—
Plateletcrit	11.9	—
Activated partial thromboplastin time	2.7	8.6
Prothrombin time	4.0	6.8
Fibrinogen—plasma	10.7	15.8
Antithrombin III—plasma	5.2	15.3
Protein C —plasma	5.8	55.2
Protein S—plasma	5.8	63.4
Factor V—plasma	3.6	—
Factor VII —plasma	6.8	19.4
Factor VIII—plasma	4.8	19.1
Factor X—plasma	5.9	—
Plasminogen—plasma	7.7	—
Von Willebrand factor	0.001	28.3

(Grouped by type rather than alphabetically.)

Appendix 2
Quality Specifications for Precision, Bias, and Total Error Allowable

The following three tables give data on *desirable* precision (CV_A), bias (B_A), and total error allowable (TE_a) for commonly assayed analytes in

1. serum or in whole blood (indicated in parentheses),
2. urine, and
3. hematology and hemostasis.

Additional data on very rarely used analytes do exist, so before you set out to determine data on analytes not included in these three tables, please go back to the original database to assess whether the data you seek already exist.

Please remember that *desirable* quality specifications are based on these formulae:

$$CV_A < 0.50\, CV_I$$
$$B_A < 0.250\,(CV_A^2 + CV_G^2)^{1/2}$$
$$TE_a < 0.250\,(CV_A^2 + CV_G^2)^{1/2} + 1.65(0.50\, CV_I)$$

However, the "three-level model" allows for analytes for which these general quality specifications cannot be met with current methodology and technology—for these difficult analyses, such as sodium, chloride and calcium, the *minimum* quality specifications are as follows:

$$CV_A < 0.75\, CV_I$$
$$B_A < 0.375\,(CV_A^2 + CV_G^2)^{1/2}$$
$$TE_a < 0.375\,(CV_A^2 + CV_G^2)^{1/2} + 1.65(0.75\, CV_I)$$

The *optimum* quality specifications for analytes for which the general quality specifications can easily be met with current methodology and technology (such as triglycerides, thyrotropin and C-reactive protein) are

$$CV_A < 0.25\, CV_I$$
$$B_A < 0.125\,(CV_A^2 + CV_G^2)^{1/2}$$
$$TE_a < 0.125\,(CV_A^2 + CV_G^2)^{1/2} + 1.65(0.25\, CV_I)$$

I encourage you to calculate these quality specifications for yourself when appropriate from the data in Appendix 1.

Table A2.1 Desirable Quality Specifications for Analytes in Serum or Whole Blood

Analyte	CV_A (%)	B_A (%)	TE_a (%)
Alanine aminotransferase	12.2	2.0	32.1
Albumin	1.6	1.3	3.9
Aldosterone	14.7	12.4	36.7
Alkaline phosphatase	3.2	6.4	11.7
Amylase	4.8	7.8	15.7
Androstendione	5.8	13.1	22.6
Angiotensin converting enzyme (ACE)	6.3	7.6	17.9
Apolipoprotein-A1	3.3	3.7	9.1
Apolipoprotein-B	3.5	6.0	11.6
Ascorbic acid	12.9	8.6	29.9
Aspartate aminotransferase	6.0	5.4	15.2
β_2-microglobulin	3.0	4.1	9.0
β-carotene	18.0	13.3	43.0
Bilirubin (total)	12.8	10.0	31.1
Bilirubin (conjugated)	18.4	14.2	44.5
CA-125 antigen	6.8	12.1	23.3
CA-15.3 antigen	2.9	10.8	15.5
CA-19.9 antigen	12.3	24.0	44.3
CA-549 antigen	4.6	8.7	16.2
Calcium	1.0	0.8	2.4
Carbohydrate deficient transferrin	3.6	9.8	15.7
Carcinoembryonic antigen (CEA)	4.7	14.1	21.8
Ceruloplasmin	2.9	3.1	7.8
Chloride	0.6	0.5	1.5
Cholesterol	3.0	4.1	9.0
Cholinesterase activity	2.7	2.9	7.4
C3 complement	2.6	4.1	8.4
C4 complement	4.5	8.6	16.0
Copper (serum)	2.5	3.6	7.7
Cortisol	10.5	12.5	29.8
C-peptide	4.7	4.1	11.7
C-reactive protein (CRP)	26.3	24.9	68.3
Creatine kinase	11.4	11.5	60.3
Creatine kinase MB, %	3.5	10.8	16.5
Creatine kinase MB, activity	9.9	7.8	24.1
Creatine kinase MB, mass	9.2	16.0	31.2
Creatinine	2.2	3.4	6.9
Dehydroepiandrosterone sulfate (DHEAS)	1.7	7.5	10.4
Estradiol	11.3	8.3	27.0
Ferritin	7.5	5.0	17.3
Follicle-stimulating hormone (FSH)	5.1	8.4	16.7
Free thyroxine (FT4)	3.8	3.6	9.9
Free tri-iodothyronine (FT3)	4.0	N/A	N/A
Fructosamine	1.7	1.7	4.5
γ-Glutamyltransferase	6.9	10.8	22.2
Globulins (total)	2.8	3.5	8.0
Glucose	3.3	2.3	7.9
Glycated hemoglobin (blood)	2.8	N/A	N/A
Haptoglobin	10.2	10.4	27.3
HDL-cholesterol	3.6	5.2	11.1
Homocysteine	3.9	7.7	14.1
Immunoglobulin A	2.5	9.3	13.4

Table A2.1 Desirable Quality Specifications for Analytes in Serum or Whole Blood (*Continued*)

Analyte	CV$_A$ (%)	B$_A$ (%)	TE$_a$ (%)
Immunoglobulin G	2.3	4.3	8.0
Immunoglobulin M	3.0	11.9	16.8
Immunoglobulin κ-chain	2.4	4.0	8.0
Immunoglobulin λ-chain	2.4	4.7	8.6
Insulin	10.6	15.5	32.9
Iron	13.3	8.8	30.7
Lactate (blood)	13.6	8.0	30.4
Lactate dehydrogenase	4.3	4.3	11.4
Lactate dehydrogenase 1	3.2	3.0	8.2
Lactate dehydrogenase 2	2.5	1.6	5.7
Lactate dehydrogenase 3	2.4	1.8	5.8
Lactate dehydrogenase 4	4.7	3.3	11.0
Lactate dehydrogenase 5	6.2	4.6	14.8
LDL-cholesterol	4.2	6.8	13.6
Lipase	11.6	10.1	29.1
Lipoprotein (a)	4.3	21.6	28.6
Luteinizing hormone (LH)	7.3	7.8	19.8
Magnesium	1.8	1.8	4.8
Mucinous carcinoma-associated antigen (MCA)	5.1	10.1	18.5
Myoglobin	7.0	8.2	19.6
Osmolality	0.7	0.4	1.5
Osteocalcin	3.2	6.0	11.2
pCO2 (blood)	2.4	1.8	5.7
pH (blood)	1.8	1.0	3.9
Phosphate	4.3	3.2	10.2
Phospholipids	3.3	3.2	8.6
Potassium	2.4	1.8	5.8
Prealbumin	5.5	2.5	14.5
Prolactin (men)	3.5	15.4	21.1
Proteins (total)	1.4	1.2	3.4
Prostate-specific antigen (PSA)	7.0	18.4	30.0
Pyruvate (blood)	7.6	5.0	17.5
Rheumatoid factor (RF)	4.3	6.5	13.5
Sex-hormone-binding globulin (SHBG)	6.1	11.1	21.1
Sodium	0.4	0.3	0.9
T3-uptake	2.3	1.6	5.3
Testosterone	4.4	5.8	13.0
Thyroglobulin	6.5	7.0	17.8
Tissue polypeptide antigen (TPA)	14.4	12.4	36.1
Tissue polypeptide specific antigen (TPS)	18.1	28.5	58.3
Thyrotropin (TSH)	9.9	8.4	24.6
Thyroxin-binding globulin (TBG)	3.0	2.1	7.1
Thyroxine (total T4)	3.0	3.4	8.3
Transferrin	1.5	1.3	3.8
Triglycerides	10.5	10.7	27.9
Tri-iodothyronine (TT3)	4.4	4.2	11.4
Urate	4.3	4.8	11.9
Urea	6.2	5.5	15.7
Zinc (serum)	4.7	3.3	11.0

CV$_G$ for free T3 used in calculation taken from Clin Chem 1988;32:962–6.

N/A—data not available

Table A2.2 Desirable Quality Specifications for Analytes in Urine

Analyte	CV$_A$ (%)	B$_A$ (%)	TE$_a$ (%)
α_1-Microglobulin, concentration, overnight	16.5	16.7	43.9
α_2-Microglobulin output, overnight	14.5	10.8	34.7
Albumin, concentration, first morning	18.0	16.4	46.1
Aldosterone, concentration	16.3	12.7	39.6
Ammonia, output	12.4	9.2	29.6
Amylase activity, random	47.0	26.2	100
Calcium, concentration	13.8	11.4	34.1
Catecholamines (total), concentration	12.0	10.0	29.8
C-Telopeptide type I collagen/creatinine, first urine	17.6	N/A	N/A
Creatinine, concentration	12.0	8.6	28.4
Deoxypyridinoline/creatinine	7.4	5.3	17.4
Estradiol	15.2	N/A	N/A
Hydroxyindolacetic acid, concentration, 24h	10.2	9.7	26.5
Hydroxyproline/minute, night urine	18.1	13.2	43.0
Magnesium, concentration	22.7	14.7	52.2
N-Acetylglucosaminidase, concentration	26.3	15.6	58.9
Nitrogen, output	7.0	7.0	18.4
N-telopeptide type I collagen /creatinine, first urine	11.6	N/A	N/A
Oxalate, concentration	22.0	11.9	48.2
Phosphate, concentration	13.2	9.4	31.1
Potassium, concentration	13.6	8.9	31.3
Protein, concentration	19.8	10.9	43.5
Sodium, concentration	12.0	9.0	28.8
Urate, concentration	12.4	8.3	28.7
Urea, concentration	11.4	8.6	27.3
VMA, concentration	11.1	13.0	31.3

Concentration data are expressed as units/volume. Output data are expressed as units/24h

The table above contains very few quality specifications for analytes as output because we usually measure the concentration of analytes and then multiply by the volume passed over time to give the output. Output, therefore, is not therefore directly measured, making it not really relevant to quote quality specifications for urine quantities in terms of output. The biological variation of analytes expressed in different ways is important because it provides information on the best way to report results (see Chapter 5).

QUALITY SPECIFICATIONS FOR PRECISION, BIAS, AND TOTAL ERROR ALLOWABLE

Table A2.3 Desirable Quality Specifications for Analytes in Hematology and Hemostasis

Analyte	CV_A (%)	B_A (%)	TE_a (%)
Hemoglobin	1.4	1.8	4.1
Hematocrit	1.4	1.7	4.1
Mean corpuscular hemoglobin	0.8	1.4	2.7
Mean corpuscular hemoglobin concentration	0.9	0.8	2.2
Mean corpuscular volume	0.7	1.2	2.3
Red cell distribution width	1.8	1.7	4.6
Erythrocyte count	1.6	1.7	4.4
Leukocyte count	5.3	5.6	14.6
Neutrophil, count	8.1	9.1	22.4
Monocyte count	8.9	13.2	27.9
Eosinophil count	10.5	19.8	37.1
Basophil count	14.0	15.4	38.5
Platelet count	4.6	5.9	13.4
Mean platelet volume	2.2	2.3	5.8
Platelet distribution width	1.4	N/A	N/A
Plateletcrit	6.0	N/A	N/A
Activated partial thromboplastin time	1.4	2.3	4.5
Prothrombin time	2.0	2.0	5.3
Fibrinogen—plasma	5.4	4.8	13.6
Antithrombin III—plasma	2.6	4.0	8.3
Protein C—plasma	2.9	13.9	18.7
Protein S—plasma	2.9	15.9	20.7
Factor V—plasma	1.8	N/A	N/A
Factor VII—plasma	3.4	5.1	10.7
Factor VIII—plasma	2.4	4.9	8.9
Factor X—plasma	3.0	N/A	N/A
Plasminogen—plasma	3.9	N/A	N/A
Von Willebrand factor	< 0.1	7.1	7.1

(Grouped by type rather than alphabetically.)

Index

A posteriori approach to reference values, 93
A priori approach
　biological variation and, 19
　reference values and, 93
Addition of variances, 21, 22, 45, 70, 73
Age
　effects on reference values, 10
　stratification according to, 97
Allowable difference between methods
　quality specifications for, 57
Altitude
　effects of, 3
American Diabetes Association
　quality specifications from, 40
Analysis of variance, 22
Analysis under optimal conditions, 20
Analytical goals, 29
Analytical performance characteristics, 29
Analytical performance goals, 29
Analytical variation
　control of in production of reference values, 95
　in production of biological variation data, 20
Anderson-Darling test, 97
Anticoagulant
　effect of, 5, 98

Best sample to collect
　biological variation and, 119
Best way to report results
　biological variation and, 119
Between-subject biological variation
　definition, 9, 16
　generation of data on, 22

Bias
　desirable quality specifications for, 54
　definition, 5
　effect of calibration on, 7
　effect on reference values, 52
　effect on serial changes in results, 75
　elimination of, 7, 57
　estimates of, 7
　false positives, false negatives and, 52
　minimum quality specifications for, 55
　optimum quality specifications for, 55
　quality specifications based on biology for, 54
　random, 8, 75
Bilirubin
　Gilbert's syndrome and, 1
　reference values and, 3
　storage of samples for, 4
Biological cycles, 10
Biological rhythms, 10
Biological variation
　best sample to collect and, 119
　best way to report results and, 119
　choosing the best test and, 120
　databases, 24
　epidemiology and, 118
　finding data on, 25
　generation of data on, 18
　method evaluation and, 121
　new methods and, 121
　number of samples and, 118
　quality specifications from, 44
　reliability coefficient and, 117

145

Calcium
  seasonal effect on metabolism, 14
Calibration
  effect on bias, 8
Case-finding
  definition, 44
  effect of individuality on, 108
Centrifugation, 5
Circadian rhythms, 10
CLIA '88
  quality specifications from, 42
Clinical fixed limits, 68
Cochran test, 22
Coefficient of variation
  formula, 4
Consensus document
  creation of, 41
Creatinine
  in men, 16
  in the elderly, 102
  in urine, 119
  in women, 16
Creatinine clearance, 121
Critical difference, 73
Critical systematic error
  definition, 62
  interpretation, 63
  in assessment of methods, 63, 125
Cycles
  daily, 10
  monthly, 12
  seasonal, 13
Cystatin C, 120

Daily variation, 10
Diagnosis
  definition, 44
  effect of individuality on, 107
Delta check failure
  algorithm for, 87,
Delta check values
  biological variation and, 87
  derivation of, 86
  LIMS and, 88

  reference change values as, 86
Desirable quality specifications
  for bias, 54
  for precision, 51
  for total error, 55
  from vignette studies, 40
Desirable standards, 29
Drugs
  quality specifications for, 58
Dynamic function tests, 11

Endogenous factors
  influence on reference values, 97
Epidemiology
  reliability coefficient in, 117
Ethnic factors
  influence on reference values, 97
European Working Groups
  quality specifications for PT and EQAS from, 58
  quality specifications for reference methods from, 59
  quality specifications from, 41
Evaluation of methods
  biological variation and, 121
Exercise
  effects of, 3
Exogenous factors
  influence on reference values, 97
Expert groups
  quality specifications from, 40
  strategy to generate quality specifications from, 41
Expert individuals
  quality specifications from, 41
External quality assessment schemes [EQAS]
  quality specifications for, 58
  quality specifications from, 42

F-test, 21
Fasting
  effects of, 3

INDEX

Fixed criteria for interpretation of results, 67
Flagging reports, 85
Further study and research, 127

Gaussian distribution
  characteristics of, 6, 47, 72
Graphical analysis to detect outliers, 23, 96
Graphic approaches to use of RCV, 84
Group biological variation, 53

Heavy chains, 108
Heavy/light chain ratio, 108
Hematological analytes
  individuality of, 109
Heterogeneity of within-subject biological variation, 22, 82
Hierarchy
  of approaches to generating reference values, 100
  of models to set quality specifications, 33
Homeostatic model, 88
Homeostatic setting point
  number of samples needed to define, 118

Immunoglobulins
  individuality of, 110
Index of individuality
  calculation of, 104
  diagnosis and, 107
  examples of in chemistry, 105
  examples of in hematology, 111
  screening and, 108
Individuality
  biological, 16, 104
  case-finding and, 17
  in hematology, 109
  reference values and, 105
  repeating tests in the healthy and, 111
  repeating test in the ill and, 112

stratification of reference values and, 107
Inter-individual biological variation, 9
International Federation of Clinical Chemistry
  recommendations on reference values and, 2
Intra-individual biological variation, 9
Introduction of methods
  biological variation and, 121
  quality specifications in, 30
Iron
  individuality of, 105

Kappa/lambda ratio
  individuality of, 108
Kolmogorov-Smirnov test, 97

Laboratory factors
  influence on reference values, 98
Light chains, 107
Linear regression transfer of reference values, 102
LIMS
  delta checks and, 88

Method decision chart, 60
  modified, 61, 125
Microalbumin, 119
Minimum quality specifications
  for bias, 55
  for precision, 51
  for total error, 55
  from vignette studies, 40
Monitoring
  definition, 44
  fixed limits and, 67
  reference change values in, 74
Monthly variation, 12

National Academy of Clinical Biochemistry
  quality specifications from, 40

National Cholesterol Education Program
  quality specifications from, 40
National Committee for Clinical Laboratory Standards
  guidelines on reference values, 91
  methods for transfer of reference values, 101
Number of samples to collect, 118
Number of subjects
  for studies on biological variation, 19
  for studies of reference values, 94
Nycthemeral rhythms, 10

Optimum quality specifications
  for bias, 55
  for precision, 51
  for total error, 56
  from vignette studies, 40
Outliers
  detection using Reed's criterion, 22, 96
  detection using the Cochran test, 21
  graphic analysis and, 23, 96

Partitioning of reference values, 9, 98
Performance characteristics, 29
Posture
  effects of, 3
Practicability characteristics, 30
Pre-analytical variation
  minimizing, 11, 19, 70, 95
  preparation for sampling and, 2
  sample collection and, 2
Precision
  desirable quality specifications for, 50
  definition, 3
  effect of analytical conditions on, 4
  effect on test result variability, 46
  effect on serial changes in results, 79
  minimum quality specifications for, 51
  optimum conditions, 20
  optimum quality specifications for, 51
  probability of change and, 80
  quality specifications based on biology for, 50
  quality specifications from vignette studies and, 39
  reference change values and, 79
Preparation for sampling
  for biological variation studies, 19
  for reference values, 95
Probability
  that change is significant, 77
  semantics and, 77
  Z scores and, 72
Proficiency testing
  quality specifications for, 59
  quality specifications from, 42

Quality assurance, 29
Quality control, 29
  calibration and, 8
Quality goals, 29
Quality improvement, 29
Quality management, 30
  six sigma and, 62
Quality planning, 32, 42, 50, 64
  introduction of, 125
Quality specifications
  based on general clinical uses, 38
  consensus conference on setting, 33
  deciding whether met, 60
  definition, 29
  difficulties in setting, 32
  for allowable difference between methods, 57
  for bias based on biological variation, 53
  for drugs, 58
  for precision based on biological variation, 50
  for reference methods, 59
  for total error allowable, 55
  for use in EQAS, 59

INDEX

for use in PT, 59
from biological variation, 44
from clinical outcome studies, 36
from EQAS fixed limits, 42
from expert groups, 41
from professional guidelines, 40
from PT requirements, 42
from the state of the art, 43
from vignette studies, 38
hierarchy of models for, 33
in action, 60
in assessing available systems, 31
in assessment of evaluation data, 31
in creating a short list, 31
in documenting requirements, 31
in introducing new analytical systems, 30
in preparing a specification, 31
in quality planning, 64
opinions on, 29
problems with, 32
strategies if not attained, 63
Quality standards, 29

Random analytical variation, 3
  sources of, 4
Random bias, 8, 75
Random walk model, 88
Reed's criterion, 22, 96
Reference change values
  as delta checks, 86
  calculation of, 73, 126
  changes in bias and, 75
  flagging reports using, 86
  for liver function tests, 76
  formula for, 73
  graphic approaches for, 84
  in different clinical units, 81
  in laboratory handbooks/user guides, 82
  in practice, 82
  in the ill, 81
  introduction of, 86
  precision and, 79
  probability and, 80
  problems with, 81
  reporting of results and, 84
Reference distribution
  definition, 92

Reference individuals
  definition, 92
  number required, 94
  preparation for specimen collection, 94
  selection for studies on biological variation, 18
  selection for studies on reference values, 94
Reference interval
  calculation of, 96
  definition, 92
Reference limits
  calculation of, 96
  definition, 92
Reference methods
  quality specifications for, 59
Reference population
  definition, 92
Reference sample group
  definition, 92
Reference values
  concept, 91
  factors affecting, 97
  generation of, 93
  hierarchy of methods for generation, 100, 126
  in cycles, 10
  individuality and, 105
  influence of bias on, 52
  influence of performance on, 52
  partitioning of, 98
  problems with, 99
  reassessment of, 126
  season and, 13
  stratification of, 98
  time of day and, 10
  time of month and, 12

Reference values *(continued)*
   transfer of, 101
   utility of, 99
Reliability characteristics, 30
Reliability coefficient
   calculation of, 117
   definition, 117
Repeating tests
   in the healthy, 111
   in the ill, 111
Replicate analyses, 71, 118
Reporting of results
   reference change values and, 84

Sample
   collection for studies on biological variation, 19
   collection for studies on reference values, 95
   handling for studies on biological variation, 19
   handling for studies on reference values, 95
   storage for studies on biological variation, 20
Sample type
   effect of, 3, 98
Screening
   definition, 44
   effect of individuality on, 108
Seasonal variation, 13
Semantics and probability, 78
Serial results
   probability of change and, 77
   reference change values and, 76
Sex
   effects on reference values, 97
   stratification according to, 98
Six sigma quality management, 62
Standard deviation
   calculation from paired samples, 21
   definition, 4
   Gaussian distributions and, 6, 48, 72
Standard normal deviate, 72

Standard operating procedures, 71, 124
State of the art
   quality specifications from, 43
Statistical factors
   reference values and, 98
Starvation
   effects of, 3
Stimulants
   biological variation and, 19
   effect of, 3
   reference values and, 94
Storage of samples
   effect of, 5
   for studies on biological variation, 20
Stratification of reference values, 9, 98
Subject
   selection for studies on biological variation, 18
   selection for studies on reference values, 93
Systematic analytical variation
   sources of, 7

Test results
   uses of, 44, 67
   fixed limits for interpretation of, 67
Test result variability
   effect of precision of, 46
Time series analysis, 88
Total error
   calculation of, 35
   definition, 34
   desirable quality specifications for, 55
   minimum quality specifications for, 55
   optimum quality specifications for, 55
   quality specifications based on biology for, 55
Total quality management, 29
Total variation
   calculation of, 22, 45, 70

INDEX

Tourniquet
  effect of, 5
Transport time
  effect of, 5
True value, 5

Variance
  definition, 21
  addition of, 22, 45, 70
Vignette studies
  design, 39
  quality specifications from, 40
Vitamin D
  effect of season on, 14

Within-subject biological variation
  constancy of, 24
  definition, 9, 16
  heterogeneity of, 22, 82
  generation of data on, 18
  in the elderly, 25
  in urine, 25
Z-scores
  calculation of total error and, 35
  probability and, 39, 78
  table of, 72